PROTECTING PERSONNEL AT HAZARDOUS WASTE SITES

PROTECTING PERSONNEL AT HAZARDOUS WASTE SITES

Edited by

Steven P. Levine
Professor of Industrial Hygiene
School of Public Health
University of Michigan

William F. Martin
Associate Director for Technology Operations: DSDTT
National Institute for
Occupational Safety and Health

With a Foreword by
David Weitzman
Manager of Industrial Hygiene Programs
U.S. Environmental Protection Agency

BUTTERWORTH PUBLISHERS
Boston · London
Sydney · Wellington · Durban · Toronto
An Ann Arbor Science Book

Ann Arbor Science is an imprint of Butterworth Publishers

Library of Congress Cataloging in Publication Data
Main entry under title:

Protecting personnel at hazardous waste sites.

"An Ann Arbor Science book."
Includes index.
1. Hazardous waste sites—Safety measures—
United States.   2. Hazardous waste sites—Hygienic
aspects—United States.   I. Levine, Steven P.
II. Martin, William F.
TD811.5.P75   1985   363.1 '79   84-19967
ISBN 0-250-40642-X

Butterworth Publishers
80 Montvale Avenue
Stoneham, MA 02180

10 9 8 7 6 5 4 3 2 1

Printed in the United States of America

CONTENTS

FOREWORD

If you can identify with these situations, help has arrived.

●You are responsible for industrial hygiene at a chemical plant and have just learned that your employees will clean up the plant's abandoned chemical dump site.

●Your professor has just gotten a State grant under which you will study industrial hygiene at a hazardous waste site cleanup operation as your thesis.

●After practicing industrial hygiene in a steel mill for fifteen years, you want to learn enough about industrial hygiene for hazardous waste site cleanup to land a job in this growing field.

These professionals need to get up to speed in industrial hygiene for hazardous waste site cleanup to meet their objectives. They also need a comprehensive reference to sustain them through the nitty gritty tasks they will face in helping to protect the workers who are working to protect our environment. Steven Levine and William Martin recognized these needs and pressed the other pioneers in this field to collaborate to produce *Protecting Personnel at Hazardous Waste Sites.*

Of course, even this excellent text does not have all the answers. But in your hands, the hands of a trained and dedicated professional, *Protecting Personnel at Hazardous Waste Sites* can be a powerful tool for assuring a healthy work environment for those individuals responsible for cleaning up this nation's numerous hazardous waste sites. The work practices and occupational safety and health programs presented in this text can be applied to current hazardous waste disposal operations as well as correcting some of our old problems due to improperly managed hazardous waste chemicals.

ACKNOWLEDGMENTS

The editors would like to thank the authors of the chapters, whose energy, and professionalism and scholarly efforts have made this book possible.

In addition, the editors would like to thank Professors Isadore Bernstein and Lawrence Fine who provided the environment and counsel necessary for creative work at the University of Michigan, and to Barbara Levine, for her support and very helpful suggestions.

Last but not least, the editors and authors owe a true debt of gratitude to the skill, patience, and dedication of both Ann Ritchey (NIOSH) who proofread, typed, organized and assembled this text, and Mary Weed (University of Michigan) who drew, and photographed the figures in this text.

THE EDITORS

Dr. Steven Levine is an Associate Professor of Industrial Hygiene at the University of Michigan, School of Public Health. Prior to his present position, Dr. Levine held positions on the Research Staffs of Stauffer Chemical and of Ford Motor Company, as well as having worked with O.H. Materials Company in the field of hazardous waste site and contaminated structure evaluation and remediation. Dr. Levine earned his doctoral degree in Analytical Chemistry from the University of Colorado in 1972. He is the author of numerous publications on the subjects of analytical and environmental chemistry.

William Fayette Martin holds a Masters degree in Environmental Health Engineering from the University of Texas, and has been a Commissioned Officer of the U.S. Public Health Service for over 20 years. He has held positions with the Indian Health Service, U.S. Coast Guard, Federal Water Pollution Control Administration, and National Institute for Occupational Safety and Health. A registered professional engineer in Texas and Kentucky, he has presented and published numerous technical papers both foreign and domestic. He served as the chairman of the Superfund steering committee made up of EPA, OSHA, NIOSH, and the U.S. Coast Guard. He served as the NIOSH Hazardous Waste Program Director with primary responsibility for coordinating all Institute Superfund activities including research projects and the production of comprehensive health and safety guidelines, worker bulletins and training materials.

THE CONTRIBUTORS

Barrett Eckert Benson has been with the USEPA National Enforcement Investigations Center in Denver, Colorado, since 1972, except for 1981 when he was an associate in the Denver office of Fred C. Hart Associates. Before joining the NEIC, he was the Chief, Division of Industrial Waste, Pollution Control Department, city of Kansas City, Missouri, and a sanitary engineer with the U.S. Public Health Service, Arctic Health Research Center in Alaska. He is a registered engineer in Missouri and Colorado and has published technical papers in ASCE, AWWA, WPCF, and other technical journals. Currently, he serves as a principal sanitary engineer for the NEIC Technical Evaluation Staff with responsibilities in the areas of water pollution and hazardous waste management. In 1980, he was awarded the EPA Silver Medal for developing the EPA methods for conducting hazardous waste site investigations.

Richard J. Costello is a senior research industrial hygienist for NIOSH, where he is responsible for occupational health effects research and development of environmental instrumentation and monitoring techniques. During the past 5 years, he has performed occupational health studies at the Chemical Control Corporation site in Elizabeth, New Jersey, at operational landfills, at hazardous waste treatment facilities and at Superfund cleanup sites. He is currently studying the occupational exposures of workers who clean up CERCLA sites under the provisions of an Interagency Agreement between NIOSH and the U.S. Environmental Protection Agency. He has authored numerous Health Hazard Evaluation Determination Reports, scientific articles and presentations and he has served as an expert witness for the Justice Department. Previously, he managed environmental and occupational health programs for the U.S. Air Force.

Mr. Costello is a member of the Waste Disposal Committee of the ACGIH and the Hazardous Waste Committee of the AIHA. He received a B.S. in Civil Engineering from the University of Arizona and a M.S. in Environmental Health Engineering from the University of Texas at Austin. He is a registered professional engineer, is certified by the ABIH, and is a Diplomate of the AAEE.

David L. Dahlstrom is Corporate Director of Safety and Health for Ecology and Environment, Inc., a firm specializing in environmental evaluation and design especially as applied to hazardous materials and waste site investigation, remediation, and personnel health and safety. Previously, he developed and presented several training programs for the USEPA and U.S. Coast Guard covering response to hazardous materials incidents, including: personnel protection and safety; field monitoring and analysis; incident mitigation and treatment; hazard evaluations; and damage assessment. He has had over 10 years experience in chemistry, microbiology, and occupational safety and health. This includes developing and managing E.E.'s employee medical monitoring, respiratory training, and field operations programs, and assisting in the design of similar programs for both governmental and industrial clients. He is the author of numerous papers on employee medical surveillance, personnel protection, health and safety program development, and hazardous materials incident response techniques.

Ralph F. Goldman has a B.S. degree in Chemistry from the University of Denver, and M.S. and Ph.D degrees in Physiology from Boston University. There, he worked on resistance to stress through endocrine mechanisms following irradiation of the adrenal glands in animals. He then worked for the U.S. Army at the Natick, Massachusetts, Quartermaster R/D Laboratories. Dr. Goldman then became the Director of a program he established in Military Ergonomics in the Army Institute of Environmental Medicine. He received numerous awards including the highest medal the Army can give to a civilian, and was appointed to the Senior Executive Service. He resigned in 1982 to help form Multi-Tech Corporation and as Senior Vice President and Chief Scientist, he is continuing his work on evaluation of human tolerance limits to heat and cold. He is working to extend human hot and cold tolerance limits by modified clothing, equipment and physical conditioning. Prediction models which accurately projects the physiological responses of workers wearing a given clothing ensemble in any work setting are current projects. He is Chairman of the NATO Research Study Group-7 on Biomedical Effects of Clothing, and has faculty appointments at Boston University, MIT and the University of Rhode Island.

E. Robinson Hoyle is currently employed as an industrial hygienist in the Occupational Safety and Health Unit of Arthur D. Little, Inc., Cambridge, MA. Mr. Hoyle was previously employed by Nelco Chemical Co., Chicago, IL, as an industrial hygienist. Mr. Hoyle has a M.S.P.H. from the University of Illinois and a B.A. from Yale University. He is the author of numerous publications and presentations in the field of industrial hygiene and has extensive experience in safety and health at hazardous waste sites.

John M. Lippitt is a Registered Sanitarian with the Ohio State Board of Sanitation Registration. He is currently employed as a Project Scientist for SCS Engineers, a consulting engineering firm specializing in hazardous and solid waste management. Mr. Lippitt provides expertise in health and safety management for SCS projects and has prepared several documents concerning methods of worker protection and costs of worker safety and health for NIOSH and the USEPA. His professional experience prior to joining SCS involved five years as a Public Health Sanitarian, a year conducting carcinogen testing research and development with the USEPA Health Effects Research Laboratory, and nine months as an on-site coordinator for the Ohio EPA to monitor the activities of a licensed hazardous waste landfill.

James Melius received his A.B. from Brown University in 1970, followed by a M.M.S. in 1972. In 1974, he received his M.D. from the University of Illinois in Chicago, followed by a residency in family practice and occupational medicine. He received his Dr. P.H. degree in Epidemiology in 1984 from the University of Illinois School of Public Health. He is board certified in Family Practice and Occupational Medicine.

Since 1980, he has been Chief of the NIOSH Health Hazard Evaluation Progam, Cincinnati, Ohio. This program conducts approximately 500 occupational health field evaluations each year throughout the country. For the past two years, Dr. Melius has been involved in evaluations at hazardous waste cleanup sites and in advising EPA on occupational health matters related to Superfund. His other research interests include occupational health problems for fire fighters, PCB combustion products, neurotoxicity, and indoor air quality.

J. Larry Payne received his B.S. degree in 1972 and his M.A. degree in 1975 from Texas A&M University, with Summa Cum Laude honors. He has extensive training experience in oil spill and hazardous material control from 1972 to the present. He has served in both graduate and undergraduate teaching positions at Texas A&M University plus work experience with Mobil Oil Research and Development Labs. He holds Fire Service Instructor and academic teaching certifications with the State of Texas. Mr. Payne has numerous honors and awards for teaching and academic achievement. He is active in several professional organizations with committee work for IFSTA Manual Validation Committee, Hazardous Materials; Technical Paper Review Committee, 1979; National Oil Spill Control Conference; and Textbook Evaluation Committee, Texas A&M University. He has been involved in over a dozen specialized training courses and conferences on Hazardous Materials over the past fifteen years.

Timothy G. Prothero has had extensive field experience performing remedial investigations and cleanups of several abandoned hazardous waste dump sites. Mr. Prothero has both planning experience and the practical "hands-on" experience of implementing those plans. His responsibilities and duties ranged from initial site investigations, remedial action planning, health and safety planning and reviews, to plan implementations, waste handling and direction of site cleanup activities. Mr. Prothero participated in and directed activities at several Superfund sites including Chem-Dyne, Pristine and Summit National in Ohio.

Mr. Prothero has also toured the continental United States on behalf of USEPA to instruct Federal, State and Local government officials on the hazards of abandoned chemical wastes, the methods and techniques used for control of those hazards, and ultimately, the proper cleanup of the orphaned sites.

Mr. Prothero has been an independent consultant since 1980, and his clients have included Federal and State agencies, and several consulting engineering firms.

Mark A. Puskar earned a B.S. degree in chemistry from Central Michigan University. In January 1981, he began working at hazardous waste sites for O.H. Materials, Findlay, Ohio, as an on-site field chemist. He took part in numerous remedial action programs. The majority of his time was spent performing instrumental analysis and compatibility testing. In January 1983, he began graduate work at the University of Michigan and was awarded an American Industrial Hygiene Foundation fellowship for the 1983–1984 academic year. He received his M.S. in Industrial Hygiene in April 1984 and is currently working towards his Ph.D. in Environmental and Industrial Health. The focus of his research is developing a FTIR–AIR method for compatibility testing.

Charles J. Sawyer is Manager, Environmental Affairs for Syntex Inc., a west coast based pharmaceutical manufacturer. His responsibilities include management of domestic and international programs of environmental engineering, industrial hygiene, and safety/fire protection. He is a registered professional engineer (Ohio and California), and a certified industrial hygienist (comprehensive practice). Prior to working at Syntex, Mr. Sawyer spent 5 years in various aspects of environmental health and safety consulting, and before that was with Procter and Gamble for 6 years. He has a bachelors degree in chemical engineering from the University of Michigan (1965), and a masters degree in chemical engineering from the University of Toledo (1967). He is an adjunct professor at the University of California—Hayward in the Environmental Health Studies program. He has published and spoken at many professional meetings on uncontrolled hazardous waste site remediation activities dealing specifically with dioxin–contaminated wastes.

Arthur D. Schwope is manager of the Applied Polymer Science Unit at Arthur D. Little, Inc. His professional activities have focused on the study of permeation through polymeric materials including the testing, analysis and specification of protective clothing. His interest in the subject began with a program for NIOSH entitled "Development of Performance Criteria for Protective Clothing Used Against Carcinogenic Liquids." He is lead author on the recent ACGIH publication *Guidelines for the Selection of Chemical Protective Clothing.* Mr. Schwope has conducted clothing studies for the U.S. Coast

Guard, NASA, the Army, the Navy, the FDA and several commercial organizations. He is chairman of ASTM subcommittee F23.30 to develop standardized test methods for assessing protective clothing materials. Mr. Schwope did his undergraduate work at Cornell University and obtained a Master's degree from MIT, both in chemical engineering.

Rodney D. Turpin is currently Safety and Occupational Health Manager, U.S. EPA, Environmental Response Team, Edison, New Jersey. In 1973, he received a B.S. in Food Industry from Delaware Valley College, Doylestown, Pennsylvania, and in 1978, a M.S. in Environmental Science from Rutgers University, New Brunswick, New Jersey. Rod's primary responsibilities are developing/implementing hazardous waste site occupational health and safety protocols, air monitoring plans and waste compatibility tests.

Dr. Lynn P. Wallace is an Associate Professor of Civil Engineering at Brigham Young University, Provo, Utah, and is a Diplomate of the AAEE. Prior to his present position, he served with NIOSH in Cincinnati, Ohio, compiling comprehensive guidelines for the protection of workers at hazardous waste sites and with the USEPA in charge of the initial hazardous waste research activities of that agency. He earned his doctoral degree in Environmental Engineering from West Virginia University in 1970 quantifying and catagorizing hospital solid wastes, including pathogenic wastes. He is the author of several publications on the protection of workers at hazardous waste sites.

INTRODUCTION AND OVERVIEW: RECOGNITION, EVALUATION AND CONTROL AT THE HAZARDOUS WASTE SITE

Steven P. Levine, Ph.D.
School of Public Health
The University of Michigan
Ann Arbor, MI 48109-2029

The problem of protecting personnel at a hazardous waste site may be viewed as being fundamentally the same as the problems encountered at traditional workplaces. That may seem like a rash statement to those of you who have only had an introduction to hazardous waste sites through the media or even from the observation platform at a hazardous waste site remedial action project. However, when you get down from the platform, "suit up" and go on the site, and then begin to analyze what you see, you will find that the only differences between a hazardous waste site and the "traditional" chemical industry workplace are the types of processes underway.

Indeed, there is no real reason that a waste site cannot be viewed as a collection of "unit processes" [1,2]. A recent study analyzed the unit process, hazardous waste drum bulking, using traditional models and concluded that fundamental industrial hygiene principles can be used to set rational personal protection guidelines [3]. Certainly, if the traditional industrial hygiene triad of recognition, evaluation and control [4-6] can be applied to such diverse work places as metal foundries, phosphorus furnace, and pesticide intermediate plants, those principles can also be applied to hazardous waste sites. In my experience as an industrial hygienist, those "traditional" work places have been more "hazardous" than are the hazardous waste sites at which I have worked. So, if there is any message to this book, it is that there is no mystery to the problem solving process at hazardous waste sites.

To emphasize the "traditional" nature of the problem, this book has been organized into three sections: recognition, evaluation and control. We hope this book will be useful in helping to prepare you for that experience on the site, both as a textbook and as a source book for public health and engineering professionals.

HISTORICAL PERSPECTIVE

During the late 1970's, our attention as industrial hygiene professionals was turned toward the use of new analytical instruments and procedures [7,8]. These procedures could be readily applied to workplace situations in which the most probable contaminants were known from a knowledge of the unit processes and materials handled at each work station. At about that time, an awareness was growing that solid waste landfill operations might pose a threat to workers [9]. This was a new type of threat, one in which the airborne contaminants were unknown and variable in composition and concentration. Hazardous waste sites required industrial hygiene monitoring aimed more at identifying airborne contaminants than with the more traditional approach of quantitating contaminants, the identity of which could be predicted from a knowledge of the process.

The efforts in the areas of hazardous waste air monitoring and worker protection were dramatically accelerated on April 20, 1980, when a fire broke out at an inactive waste treatment facility called, ironically, "Chemical Control Corporation," in Elizabeth, New Jersey. The tens of thousands of leaking and burning tanks and drums at the site obviously posed a real problem; the firemen fighting the fire and the workers during the two year long post-fire cleanup were potentially exposed to unknown and variable sources of toxic chemicals. The following true story emphasizes the inexperience of our profession with problems at hazardous waste sites as of that date.

Dick Costello (the author of Chapter 5 in this book) was sent by NIOSH to the site of the Chemical Control fire and subsequent cleanup to assess the potential danger to workers on the site, and to help develop a personal protection strategy for those workers. Besieged by reporters, he escaped to a stall in the men's room where he finally achieved a degree of privacy necessary to think. Finding this office well stocked with paper, he wrote down the first health hazard evaluation strategy and the beginnings of a personal protection plan for hazardous waste workers.

RECOGNITION

The first NIOSH effort at Chemical Control ultimately evolved into a joint NIOSH, U.S. EPA, U.S. Coast Guard, U.S. OSHA effort [15], the genesis and products of which have been adequately summarized elsewhere [10,16]. The direction of that interagency program has been the work of Bill Martin who is the co-editor of this book and the author of Chapter 2. The fruits of the labors of these organizations have been training courses for first responders to hazardous waste emergencies, worker training bulletins, worker training courses and documents, and an occupational safety and health guidance manual [17-24]. Some of these courses and documents are still being prepared at the time of the writing of this book. All are expected to play significant roles in the personal protection strategies that are evolving. (In addition, a number of general purpose textbooks have been published that represent a good supplement to these government documents [25-28].)

Industrial programs for worker and community protection include those organized by the Chemical Transportation Emergency Center (CHEMTREC) [29,30], programs organized by individual companies [31], and those designed to ensure the health of workers at specific sites [32] using specific technologies [33,34]. Chapter 3 written by David Dahlstrom presents a summary of an industrial program. This chapter can provide the nucleus around which a good waste site health and safety plan can be drawn. Indeed, few additions or alterations are needed to use this Chapter 3 for that purpose.

The last chapter in this section is written by Barrett Benson from a perspective not usually associated with industrial hygiene activities, that of the initial site investigator. This person gathers all available data and identifies missing data on a hazardous waste site. That data can then be used to formulate priorities for ranking of sites [35-37], remedial action strategies and bid specifications [38,39], and worker [40] and community protection guidelines. This chapter, used in conjunction with Chapter 2, will provide you with a wealth of information with which these data and strategies can be accessed.

EVALUATION

In the 1970's, I frequently found it possible to rapidly evaluate the magnitude of the workplace problem without the need for extensive instrumentation. Indeed, the fires in the

3

footprints in the dust inside the plant and the pitted windshields on the cars in the parking lot from airborne fluorides at a phosphorus furnace plant were easy clues that helped in the evaluation of that workplace. When I employed instrumentation to assist in the evaluation, the results were usually quite striking. For example, once I was attempting to quantitatively evaluate dust exposure at a work station in a PVC fabricating plant and the automatic dust monitor I employed promptly read off-scale. This was to be expected since it was coated with chromate dust within minutes of starting. These examples are given because they contrast strongly with the experience of those who have performed air monitoring at hazardous waste sites.

Dick Costello of NIOSH probably has more experience to date with this type of health hazard evaluation than any other industrial hygienist [3,41,42], although other groups have published on this subject [33,43,44]. It is interesting to note that his results have shown low concentrations of airborne contaminants at hazardous waste sites. In this chapter, he recognizes the need for simple, direct measurement devices [45-47] in the evaluation of airborne hazards since contaminants at hazardous waste sites tend to be episodic, not continuous. Although there are new technologies that may emerge as continuous monitors with the power to speciate and quantitate in near-real time [48,49], these are currently only research tools. This chapter focuses instead on the strategy that must be used in the gathering and interpretation of results. This is an appropriate focus for such a chapter in an industrial hygiene book.

The evaluation of the potential for chemical hazards in the workplace, whether the hazards result from airborne exposure or direct contact, is always easier when you know what chemicals are being handled. The advantage I had in the phosphorus furnace plant was that I knew that phosphorus was being produced and that the ore contained fluoride. It is easy to look smart if you know the unit processes and the materials handled. Indeed, several useful hazard indices have been developed for that type of situation [50,51]. At hazardous waste sites such as Chemical Control, you may be able to identify the unit processses, but the real problem resides in not knowing what chemicals are being handled. This is the subject area covered in Chapter 6.

One of the chemists who is a graduate of Chemical Control and many other such sites, and who actively participated in sampling and analysis of the waste materials is Mark Puskar. His "war stories" include incidents of drums exploding near the sampling stations and of inadequate analysis techniques leading to misclassification of the

contents of drums. Rod Turpin, who has co-authored this
chapter with us, is one of the pioneers in this field. He
has many similar "war stories". Although gas chromatography
- mass spectrometry (GC-MS) may be employed to unequivocally
identify and quantitate the contents of each layer in each
drum [52], the cost and time required makes this technique
unrealistic (assume $500/assay X 40,000 assays = $20,000,000
for GC-MS assays alone at a typical site). If we are to
depend on quick, inexpensive, and reliable compatibility
testing, we must first understand the alternative approaches
and then select the methods that best meet our site-specific
needs. This is the focus of Mr. Puskar's chapter. He also
makes the point that an aggressive quality control program is
as necessary for compatibility testing as it is for any other
analysis scheme. Incidentally, the development of a new
compatibility testing procedure based on Fourier Transform
Infrared Spectroscopy (FTIR) is the focus of Mr. Puskar's
Ph.D thesis research at the University of Michigan [53].

Chapter 7 deals with the subject of medical
surveillance. The message of this chapter is that prudent
surveillance schemes should be employed in combination with
archiving blood samples for future use in case of suspected
exposure to a specific chemical. This is the approach used
in other industrial settings [54], but frequently hazardous
waste workers are needlessly subjected to every type of blood
chemistry available simply as a defensive measure. Dr.
Melius, the author of this chapter, points out that this is
not necessary. A well defined program based on understood
principles [55-57] is more desirable than administering
unwarranted tests. He also points out that advance planning
with the local hospital is very important in the case of
medical emergency.

CONTROL

As industrial hygienists, we are always taught that
engineering controls are the first line of defense for the
worker [4-6]. Those of you that need to be convinced only
need to don a one-piece Saranex-Tyvek protective garment with
integral hood and booties, tape on your gloves and
overbooties, strap on a "lightweight" 25 pound self-contained
breathing apparatus (SCBA), and pick up a 5 pound tool. Do
this on a warm summer day for about 15-30 minutes, and you
will appreciate the desirability of working to maximize
engineering controls and minimize reliance on protective
garments and breathing apparatus to protect personnel at
hazardous waste sites.

Chapter 8, which deals with engineering controls, was written by Dr. Lynn Wallace, formerly a colleague of Bill Martin's at NIOSH, and presently at the Department of Civil Engineering of Brigham Young University. He points out that, if you work for a company that low bids a contract by using low cost labor and little capital equipment, you will have to deal with protecting workers that are muscling drums around a site. On the other hand, simple things like barrel grapplers and sparkproof drum punches attached to backhoes make the job of the health and safety professional much easier. When you combine the "automated" equipment with a sensible site layout (such as a well placed and well designed barrel staging and bulking area) [58], the potential for the occurrence of conditions immediately dangerous to life and health (IDLH) [19] is greatly reduced. Since most decisions on personal protection strategies at waste sites seem to be made on the assumption that IDLH conditions may occur, the use of engineering controls to reduce that risk is highly desirable.

The author of Chapter 9, Art Schwope, virtually "wrote the book" on the efficacy of modern fabric materials for use in protective garments [59,60]. Of course, he had a little help from other groups working on this subject [61-67]. The difference is that Art was the first to characterize the impermeability of Saranex-Tyvek. This garment is one of the most widely used protective garment fabrics because of its impermeability to chlorinated organics, as well as its light weight, strength and ease of fabrication. In addition, it is disposable because of its low cost. For those of you who have tried to decontaminate, store and reuse expensive garments, you know the value of a disposable suit.

His chapter also deals with the question of respiratory protection. The question of setting criteria for determining when the use of Level B (SCBA) or Level C (full-face respirator) protection is necessary at a hazardous waste site is still unresolved. Most sites use Level B protection under all circumstances where the potential for IDLH conditions and/or the type and concentration of airborne contaminants is undefined. Although this may or may not be necessary [68,69], a conservative approach to industrial hygiene practice is usually used and SCBA are employed. This results in what I estimate to be an approximately 30-60 percent loss in efficiency because of the added weight on the worker, the need to refill air tanks, and the expense of manning and maintaining a refilling and maintenance station. Whether you wear SCBA or a full-face cartridge respirator, the problems of communication and visibility will be with you. I was surprised one morning when I removed the nose cup from my face mask to allow better communication and wound up fogging my facepiece even though the ambient temperature was 70 F.

6

Perhaps the most scholarly chapter in the book is that on heat stress. Although this subject has been addressed in a variety of books and other publications [70-77], never before has the question been explored in such depth and so coherently. The reader is advised to don their thinking cap before attempting to read this chapter. However, it is well worth the effort since the question of heat stress is perhaps the most important open issue in worker protection at hazardous waste sites. Too often we protect the workers against potential or imagined potential chemical hazards and subject them to unwarranted and debilitating heat stress. Those of you who have followed my advice to suit up and experience the site firsthand will understand this point.

Decontamination is the subject of Chapter 11 which is written by John Lippitt. Fortunately, this subject has been addressed quite well in recent remedial action bid specifications issued by the U.S. Army Corps of Engineers [58]. The use of equipment decontamination pads, boot rinses, shower and change trailers, on-site laundries, equipment buildings and water collection systems are all part of this vital question. These questions are more important than the specific layout of the "decontamination line" that was addressed in earlier documents [19]. This is especially true since there has been a major switch to the use of disposable garments. Understandably, it took time for these engineering controls and facilities to evolve, but now there is a clear expectation that sites will be operated with a full complement of such facilities.

One should not overlook the minor work practices that make a decontamination zone function effectively. Are the workers employing the technique of self-criticism to see to it they do not break the decontamination line for some thoughtless reason? Is the suit change assistant going into the food trailer to refill the ice water jug? These "small things" frequently turn out to be the most easily missed and thus very important when a decontamination protocol is put into practice.

None of the controls or plans will work to protect workers unless they are adequately trained. The chapter on training is authored by Larry Payne, a member of the faculty of the Texas A&M Oil and Hazardous Material Control Training Program, which is the oldest and perhaps best such training program in the country. This program is used to train workers and managers alike. All too often, either untrained (low salary, low tenure) workers are teamed with managers who are untrained, inexperienced and/or careless. The bottom line may be profits for the remedial action contractor and staying within budget for the governmental coordinator, but unless money is spent on training, both will come to grief.

The training must not only precede site work, but also must take place every day, preferably in morning safety meetings. At these meetings, worker self-criticism can reinforce previous training, and managers can inform workers about special hazards to be encountered that day. In addition, mini-sessions can be conducted to reinforce previous training.

The final chapter, aptly numbered Chapter 13, is the "if all else fails" chapter. The subject is contingency planning. It is written by Chuck Sawyer who, aside from being the co-founder of the AIHA Hazardous Waste Committee, supervised the cleanup of the dioxin-laden waste at Verona and Denny Farms, Missouri. Needless to say, the contingency plans required for those projects has to cover everything from broken legs to chemical contamination to tornados. The lessons learned at those sites are detailed in this chapter.

We have already discussed the importance of a multi-step approach to safety and health. The first step is information gathering and pre-planning, followed by site air and materials monitoring and evaluation. After that, engineering controls and personal protective devices must be employed along with a sound and continuing training program. The last, and very important, part of these tiers of protection is the contingency plan. When all of these steps are implemented, the health and safety of the worker is insured against all acts of men, and perhaps even some acts of God. This is the same approach that is used for worker protection at "traditional" work sites. The purpose of this book is to assist you in your industrial hygiene training in order to achieve that end.

REFERENCES

1. Cralley, L.V. and, L.F. Cralley, Eds. Industrial Hygiene Aspects of Plant Operations, Volume 1.(New York: Macmillan, 1983).

2. Shreve, R. N., and J. A. Brink, Jr. Chemical Process Industries, (New York: McGraw-Hill Book Co., 1977).

3. Levine, S. P., R. Costello, C. Geraci, and K. Conlin. "Setting Personal Protection Guidelines at a Hazardous Waste Site Unit Process," submitted to Amer. Ind. Hyg. Assoc. J., 1984.

4. Radcliffe, J. C., et al. "Industrial Hygiene Definition, Scope, Function and Organization," Ind. Hyg. Assoc. J., 1984.

5. "The Industrial Environment - Its Evaluation and Control," U.S. Department of Health, Education, and Welfare (NIOSH), 1976, pp. 3-4.

6. Olishifski, P. E., and F. E. McElroy, Eds. "Fundamentals of Industrial Hygiene," (Chicago, IL: National Safety Council, 1977) Chapter 1.

7. "Air Sampling Instruments," (Cincinnati, OH: American Conference of Government Industrial Hygienists, 1978).

8. Taylor, D. G., Ed. "NIOSH Manual of Analytical Methods," 2nd ed., U.S. Department of Health, Education, and Welfare, NIOSH (1977-1981).

9. Wilson, D. G. "Chapter 6 (Condensed from 'Handbook of Solid-Waste Management,' 1977)In Dangerous Properties of Industrial Materials, N. I. Sax, Ed. (New York: Van Nostrand, 1979).

10. Harris, L. R., and S. Howards. In "Proceedings," National Conference on Hazardous and Toxic Wastes Management, Newark, NJ, June 1980, pp. 182-191.

11. Melvold, R. W., S. C. Gibson, and M. D. Royer. In "Proceedings," Management of Uncontrolled Hazardous Waste Sites Conference, Washington, D.C., October 1981, pp. 269-276.

12. Blum, B. In "Hazardous Waste Disposal," Highland, J. H., Ed., Ann Arbor Science: Michigan, 1982, pp. 3-7.

13. Bennet, G. F., F. S. Feates, and I. Wildee. "Hazardous Materials Spills handbook," McGraw-Hill: New York, 1982, pp. 12-114.

14. Martin, J. M., L. P. Wallace, and R. F. Coene. "Abstract of Papers," Amer. Ind. Hyg. Assoc. Conf., Philadelphia, Pa., May 1983, paper no. 102.

15. Memorandum of Understanding, Interagency Agreement AD-75-F2A091. U.S. EPA and U.S. Dept. of Health and Human Services.

16. Scott, D. M. and D. M. Robinson. "The NIOSH Program for Protecting Workers Potentially Exposed to Hazardous Waste/Substances," NIOSH (DCDSD-Special Projects), Draft Document, March 1981.

17. "Hazardous Waste Sites and Hazardous Substance Emergencies," DHHS (NIOSH) Publication 83-100, December 1982.

18. "Oil and Hazardous Substances Response Manual," U.S. EPA Region 7, (undated).

19. "Hazardous Materials Incident Response Operations Training Manual," U.S. EPA Office of Emergency and Remedial Response, Hazardous Response Support Division, June 1982.

20. "Hazardous Materials Response Manual," U.S. Coast Guard, Environmental Coordination Branch, Draft Document, May 1982.

21. "EPA's Emergency Response Program," U.S. EPA, HW-3, April 1982.

22. "Environmental Response Team," U.S. EPA, HW-2, April 1982.

23. "Pocket Guide to Chemical Hazards," U.S. DHHS (NIOSH) Publication No. 78-210, 1981.

24. "Occupational Health Guidelines for Chemical Hazards," U.S. DHHS (NIOSH) Publication No. 81-123, 1981.

25. "Polychlorinated Biphenyl Destruction Technology," Dillon: Toronto, 1982.

26. "Detoxication of Hazardous Waste," Exner, J. H., Ed., Ann Arbor Science: Michigan, 1982.

27. "Hazardous Waste Management for the 80's," Sweeney, T. L., et al, Eds., Ann Arbor Science: Michigan, 1982.

28. "Toxic and Hazardous Waste Disposal," Pojasek, R. B., Ed., Ann Arbor Science: Michigan, 1982, Vol. 4.

29. Wright, C. J. "Recognizing and Identifying Hazardous Materials," Inter-Industry Task Force on the Rail Transportation of Hazardous Materials, 1979.

30. Zercher, J. D. In "Proceedings," 1982 Hazardous Materials Spills Conference, Milwaukee, Wisc., April 1982, pp. 231-237.

31. Ibid., Lusher, L. W., Jr., pp. 238-246, Niggel, W. T., pp. 247-249, Brand, S. H., pp. 250-251.

32. Exner, J. H., et al. In "Detoxication of Hazardous Waste," Exner, J. H., Ed., Ann Arbor Science: Michigan, 1982, Chapter 17.

33. Sawyer, C. J., Ibid., Chapter 18.

34. Vander Velde, G. Chemical Waste Management Co., personal communication, October 1982.

35. "Comprehensive Environmental Response, Compensation and Liability Act," Public Law 96-510, December 11, 1980.

36. "Document Records for Hazard Ranking System, Chem-Dyne," U.S. EPA Region V, August 12, 1982.

10

37. "Potential Hazardous Waste Site; Site Inspection Report, Chem-Dyne," Ecology and Environment Co., September 8, 1982.

38. Harsh, K. "Emergency Action Plan, Chem-Dyne, Hamilton, Ohio," Ohio EPA, 1981.

39. "A Hazardous Waste Site Management Plan," Chemical Manufacturers Assoc.: Washington, D.C., 1982; Appendix XII, Section 7.0.

40. "Work Safety Protocols," Chem-Dyne Contract RFQ Package, Section 1G, U.S. Army Corps of Engineers, January 1983.

41. NIOSH Health Hazard Evaluation Report HETA 81-037-1055 (Rollins Environmental Services, Baton Rouge, LA).

42. Costello, R. J., and M. V. King. "Protecting Workers Who Clean Up Hazardous Waste Sites," Amer. Ind. Hyg. Assoc. J., 43:12 (1982).

43. D'Appolonia, K. J. "Health Matrix-Toxic Waste Isolation," Amer. Ind. Hyg. Assoc. J., 43:1 (1982).

44. Harsh, K. M. In "Hazardous Waste Management for the 80's," Sweeney, T. L., et al, Eds., Ann Arbor Science: Michigan, 1983, Chapter 14.

45. Comparison of Portable Hydrocarbon Analyzers," U.S. DHEW (NIOSH) Publication No. 760166.

46. Schneider, D. and K. Leichnitz. "Drager Gas Detection Kit," and "Qualitative Detection of Substances by Means of Drager Detection Tube Polytest," Drager Review 46, 1980.

47. King, M. V., P. M. Eller, and R. J. Costello. "A Qualitative Sampling Device for Use at Hazardous Waste Sites," Amer. Ind. Hyg. Assoc. J., 1983, 44, 615.

48. Thomson, B. A., et al. "Fast In-Situ Measurement of PCB Levels in Ambient Air to the ng/m3 Level Using a Mobile APCI Mass Spectrometer System," presented at 8th Intl. Mass Spectrom. Conf., Oslo, Norway, August 1979.

49. Herget, W. F. and J. D. Brasher. "Remote Optical Sensing of Emissions," Appl. Opt., 18:3404 (1979).

50. Corn, M., and N. A. Esmen. "Workplace Exposure Zones for Classification of Employee Exposures to Physical and Chemical Agents," Amer. Ind. Hyg. Assoc. J., 40:47 (1979).

51. Langner, R. R., et al. "Two Methods for Establishing Industrial Hygiene Priorities," Amer. Ind. Hyg. Assoc. J., 40:1039 (1979).

52. Harsh, K. M. In "Proceedings," 1982 Hazardous Materials Spills Conference, Milwaukee, Wisc., April 1982, pp. 199-204.

53. Levine, S. P., M. A. Puskar, C. Geraci, M. Bolyard, R. Costello. "Rapid Identification of Hazardous Waste Drum Constituents Using FTIRATR," submitted to Amer. Ind. Hyg. Assoc. J., 1984.

54. Noonan, R., Manager, Industrial Hygiene and Environmental Engineering, Amtrak, personal communication, March 1983.

55. "Environmental and Occupational Medicine," Rom, W. N., Ed., Little, Brown: Boston, 1983.

56. "Occupational Diseases – A Guide to Their Recognition," Key, M. M., et al, Eds., U.S. DHEW (NIOSH) Publication No. 77-181, 1978.

57. Melius, J. M. and W. Halperin. "Medical Screening of Workers at Hazardous Waste Disposal Sites," presented at the Society for Occupational and Environmental Health Conference on Hazardous Waste, Washington, D.C., December 1980.

58. "Work Safety Protocols," Chem-Dyne Contract RFQ Package, U.S. Army Corps of Engineers, January 1983.

59. Schwope, A. D., et al. "Guidelines for the Selection of Chemical Protective Clothing," ACGIH, Cincinnati, Ohio, 1983.

60. Schwope, A. D. "The Effectiveness of Tyvek Composites as Barriers to Aroclor 1254, Trichlorobenzene, and Mineral Spirits," A. D. Little Co.: Massachusetts, 1982.

61. William, J. R. "Chemical Permeation of Protective Clothing," Amer. Ind. Hyg. Assoc. J., 1980, 41, 884.

62. Sansone, E. B. and Y. B. Tewari. "Permeability of Protective Clothing Materials to Benzene Vapor," Amer. Ind. Hyg. Assoc. J., 1980, 41, 170.

63. Coletta, G. C., et al. "Development of Performance Criteria for Protective Clothing Used Against Carcinogenic Liquids," U.S. DHEW (NIOSH) Publication No,. 79-106, 1978.

64. Weeks, R. W., Jr. and M. J. McLeod. "Permeation of Protective Garment Materials by Liquid Halogenated Ethanes and a Polychlorinated Biphenyl," U.S. DHHS (NIOSH) Publication No. 81-110, 1981.

65. Weeks, R. W., Jr. and M. J. McLeod. "Permeation of Protective Garment Material by Liquid Benzene and by Tritiated Water," Amer. Ind. Hyg. Assoc. J., 1982, 43, 201.

66. Williams, J. R. "Permeation of Glove Materials by Physiologically Harmful Chemicals," Amer. Ind. Hyg. Assoc. J., 1979, 40, 877.

67. Sansone, E. B. and Y. B. Tewari. "Differences in the Extent of Solvent Penetration Through Natural Rubber and Nitrile Gloves," Amer. Ind. Hyg. Assoc. J., 1980, 41, 527.

68. Mine Safety and Health Administration. "Code of Federal Regulations," 30 CFR, Subchapter B, part 11.3 (+), Respiratory Protection Apparatus, U.S. Government Printing Office, Washington, D.C., 1981.

69. Teresinski, M. F. and P. N. Cheremisinoff. "Industrial Respiratory Protection," Ann Arbor Science: MIchigan, 1983.
70. "Criteria for Recommended Standard for Hot Environments," U.S. DHEW (NIOSH) Publication No. 72-10269, 1972.
71. "Threshold Limit Values," Amer. Council of Gov. Ind. Hyg. (ACGIH): Cincinnati, OH, 1982, pp. 60-67.
72. "Engineering Field Reference Manual," Amer. Ind. Hyg. Assoc. (AIHA): Akron, 1982, Chapter 5.
73. "Fundamentals of Industrial Hygiene," Olishifski, P. E., and McElroy, F. E., Eds. National Safety Council: Chicago, 1977, Chapter 8.
74. Pasternack, A. "Cooling and Respiratory Protective Equipment," Drager Review, 1979, 44, 12-23.
75. "ILC Cool Vest 19," ILC Industries: Frederica, Del., 1983.
76. Pasternack, A. "The Ice Cooling Vest," Drager Review, 1982, 49, 1-6.
77. Pasternack, A. "Biomechanical and Work Physiological Evaluation of the Cooling and Respiratory Protective Device," Drager Review, 1981, 47, 12-17.

2

FEDERAL GOVERNMENT PROGRAMS

William F. Martin, PE
Engineer Director
Hazardous Waste Program
National Institute for Occupational Safety and Health
Cincinnati, Ohio 45226

The Federal government has historically taken an active role in providing technical assistance and disseminating information. These services are provided to the general public, academia, and state and local organizations in the form of research support, supplemental funding, training programs and information programs. Provision of these services is usually mandated by legislation, and broad public access to the materials within Federal agencies is insured by the Freedom of Information Act of 1966 [1].

The resources available on protecting worker safety and health from hazardous substances are as diverse as the many Federal programs themselves. There has been a rapid growth of available resources due to the recent emphasis in three areas: (1) research; (2) the gathering of information nationwide for identifying health risks associated with hazardous wastes; and (3) the production of publications to assist state and local organizations in the recognition, evaluation, and control of hazardous substances. Due to modern information storage, retrieval, and data base management, this vast and ever-growing body of technical information is more readily accessible to the private sector than ever before.

Even though a tremendous amount of technical data and other resources are available, much of the potential audience may never benefit from them. Some potential users will not know where they exist or how to access them, and an unfortunate few will never even know of their existence. For similar reasons, many of the users who do access Federal programs will never receive the full benefits of these programs.

This chapter will attempt to provide potential users with information and insights which will assist them in becoming full recipients of these services. It will, at best, serve as a starting point, since utilizing a Federal program requires skill, knowledge, and tenacity. Potential users must obtain a basic knowledge of the program's responsibility, purpose, funding level, eligibility requirements and delivery mechanisms. Since this information is often provided in the legislation and public law itself, the following section, Legislative Background, will highlight some of the laws which pertain to hazardous substances and worker safety and health. Responsible agencies and their functions will be identified in the section, Federal Agencies and Programs. Potential users are advised to obtain the promotional booklets developed by these agencies. They generally contain more useful information than the legislation, and they describe the program in greater depth and in the terminology of the potential user. By initially accessing these Federal programs, potential users will have already moved beyond this chapter. The sections, Accessing Federal Programs and Program Change, will attempt to provide some insight into why familiarity with the programs is necessary, and also why tenacity is absolutely essential.

LEGISLATIVE BACKGROUND

Historically, most of the occupational safety and health laws were enacted by states. Prior to 1960, there were relatively few Federal laws addressing worker protection, and those were only applicable to a limited number of employers or specific groups of employees [2]. During the 1960s, there was increased concern about worker safety and health in this nation, and the result was a proliferation of Federal legislation. Ultimately, a nationwide occupational safety and health program was designed through the Occupational Safety and Health Act of 1970 (OSHAct). The OSHAct extended health and safety coverage to most workers of business affected by interstate commerce [3].
The OSHAct established three organizations:

1. The Occupational Safety and Health Administration. Located within the Department of Labor, OSHA is primarily responsible for the promulgation and enforcement of standards and worker training.

16

2. The Occupational Safety and Health Review Committee. The OSHRC is an independent agency within the Executive Branch which adjudicates contested cases resulting from OSHA-initiated actions against employers.

3. The National Institute for Occupational Safety and Health. NIOSH is located within the Department of Health and Human Services (formerly HEW), and is administratively under the Centers for Disease Control. NIOSH is the principle Federal agency engaged in research, education and training, and disseminating information related to occupational safety and health.

Even though the OSHAct covers most businesses in the country, it was not specifically applied to hazardous waste handling until a 1982 amendment was made to the National Contingency Plan under Public Law 96-510, the Comprehensive Environmental Response, Compensation and Liability Act of 1980 (Superfund). In addition to OSHA and NIOSH, there are other Federal agencies with specifically defined responsibilities for assisting in protecting workers from adverse health effects from hazardous wastes. One such agency is the Environmental Protection Agency (EPA). There are also other agencies and programs with technical resources related to the handling and management of hazardous substances and to worker protection. Many of these programs will be identified in the section on Federal agencies.

The number of diverse Federal programs which relate to hazardous wastes is due to both the complexities and the multidisciplinary aspect of hazardous waste management and to the extensive amount of Federal legislation which addresses hazardous materials. Much of the legislation on hazardous substances which applies to worker safety and health was developed to regulate activities other than waste site management or remedial actions. The National Contingency Plan in 40 CFR 300 is a notable exception.

Among the regulations and legislation which apply to worker safety and health are the following:

* The Occupational Safety and Health Act of 1970 and the subsequent passage of regulations 29 CFR notably Parts 1910 (Occupational Safety and Health Standards) and 1926 (Safety and Health Standards for Construction).

* The Resource Conservation and Recovery Act of 1976 (RCRA) and subsequent regulations in 40 CFR 260 through 2655 such as Section 265.16 (Personnel Training) and 265 Subparts C (Preparedness and Prevention) and D (Contingency and Emergency Procedures).

17

* The Comprehensive Environmental Response, Compensation
  and Liability Act of 1980 (CERCLA or Superfund Act)
  and its regulations (i.e., the National Contingency
  Plan) under 40 CFR, Sections 300.45 (Local
  Contingency Plans) and 300.71 (Worker Safety and
  Health) which stipulate emergency planning and
  coordination and enable the lead agency or on-scene
  coordinator to specify regulations and guidelines
  which should be followed.
* The Atomic Energy Act of 1954 and subsequent
  regulations in 10 CFR 20 which stipulate standards
  for protection against radiation.
* Federal Mine Safety and Health Act (FMSHA) of 1977 and
  subsequent regulations found in 30 CFR 11 (Department
  of the Interior, Bureau of Mines, Respiratory
  Protective Devices and Tests for Permissibility)
  provide the primary technical criteria for a
  permissible respirator.
* The Hazardous Materials Transportation Act of 1975
  (HMTA) and regulations in 49 CFR Parts 100 to 199
  which stipulate labeling, marking packaging,
  placarding, manifesting and shipping papers, vehicle
  design, vehicle maintenance, and handling of
  hazardous materials during loading, transport, and
  unloading.

Potential users are strongly urged to have a cursory
familiarity with these pieces of legislation, as well as the
following additional legislation listed below, which are
related to hazardous substances and environmental
emergencies or toxic emergencies.

Toxic Substances Control Act;
Federal Insecticide, and Rodenticide Act;
Federal Disaster Relief Act;
U.S. Army Corps of Engineers Emergency Authorities, and
Consumer Product Safety Act.

There are other pieces of legislation which will be of
interest to the industrial hygienist working with hazardous
material management; however, familiarity with those listed
above will provide an adequate foundation for initially
accessing Federal programs and their resources.

FEDERAL AGENCIES AND PROGRAMS RELATING TO PROTECTION OF
HAZARDOUS WASTE WORKERS

This section will identify and highlight many of the
Federal programs which relate, directly and indirectly, to

health and safety protection of hazardous waste workers. The informational and technical resources available through these programs will be emphasized, rather than their regulatory and policy-making functions. These resources are divided into three categories: (1) information, training, and technical assistance; (2) data bases and reference volumes, and (3) hot-lines and emergency response resources. Even though an agency or a specific program may serve all three functions, this classification of resources is presented because of the distinctly different user needs and the diverse characteristics of program activities within each category.

The discussions of the programs will be brief, since the purpose of this chapter is to provide a guide for inquiry and access to services. An exception is made for the discussion of the Occupational Safety and Health Administration (OSHA) and the National Institute for Occupational Safety and Health (NIOSH). They are agencies engaged in research, education and training, and providing information related to occupational safety and health. OSHA and NIOSH are thus good sources of information and resources on both worker safety and health and on hazardous substances found in the workplace [4, 5].

Information and Technical Assistance

National Institute for Occupational Safety and Health (NIOSH)

1. Health Hazard Evaluations which provide on-site evaluation of potentially hazardous chemical, physical, and biologic exposures. HHE's have both medical and industrial hygiene components. Employers or representative groups of employees may place requests.
2. The technical information available includes NIOSH criteria documents, publications, and research reports. Criteria documents contain a review of the scientific literature, required medical controls, methods of sampling and analysis and probable safe atmospheric levels. In response to inquiries, NIOSH will search two data bases which relate to toxic substances and worker safety and health (RTECS and NIOSHTIC).
3. Technical assistance is available in the areas of epidemiology, engineering, industrial hygiene, and occupational medicine.

4. Training courses are provided by NIOSH and by colleges and universities through NIOSH contracts and grants. Courses are available in the fields of occupational medicine, occupational nursing, industrial hygiene, and occupational health engineering.
5. As stated earlier, NIOSH is administratively under the Centers for Disease Control (CDC), and inquiries to NIOSH often yield valuable environmental health information referrals to CDC's programs and specific referrals to individual professionals.

Occupational Safety and Health Administration (OSHA)

In addition to its enforcement responsibilities, OSHA provides technical assistance, publications, training courses, and inspections. OSHA also provides consultation to help businesses comply with safety standards. Most of these resources can be obtained through the ten regional offices.

Centers for Disease Control (CDC)

CDC provides assistance through its Center for Environmental Health. Technical assistance, publications, laboratory and analytic services, and disaster and emergency response are available. CDC is the focal point for the Department of Health and Human Service responsibilities under CERCLA.

The Federal Emergency Management Agency (FEMA)

FEMA participates in the development and evaluation of national, regional, and local contingency plans. It monitors responses to these plans and evaluates requests for presidential designation of disaster areas. After a presidential declaration, FEMA coordinates the various types of Federal assistance considered necessary to protect life and property including emergency evacuation. FEMA operates the National Training Center and National Fire Academy, and in cooperation with the Environmental Protection Agency (EPA), has sponsored the development of a planning guide, Preparing for Environmental Disaster and a related training course. Contact FEMA about this guide and course, and other related courses.

U.S. Geological Survey (USGS)

The USGS provides expertise on geology, hydrology, and soil. Publications and expertise are also available in the areas of oil drilling, producing, handling and transporting by pipeline. An extensive collection of topographic maps is available. These maps and many of the USGS reports may be useful in site selection and in remedial actions.

National Bureau of Standards

The Bureau provides publications on its own measurement programs in chemistry and other areas, and from its reference collection of engineering standards and specifications issued by Federal agencies, technical societies and trade associations.

Department of Energy (DOE)

The DOE operates the Technical Information Center at Oak Ridge, Tennessee, which disseminates its own reports, as well as international nuclear science literature. It also supports seven other information centers, including the Nuclear Safety Information Center. The U.S. Energy and Development Administration (ERDA) serves as coordinator for the Interagency Radiological Assistance Plan (IRAP). ERDA works with other Federal, state and local agencies in disseminating technical guidance and support for radiation emergencies. Contact should be made through ERDA regional offices.

National Oceanic and Atmosphere Administration (NOAA)

NOAA provides detailed state-of-the-art meteorologic data and scientific support including expertise on living marine resources. This organization is conducting research on several topics of hazardous substances that will assist health and safety professionals.

Department of Transportation (DOT)

The Department of Transportation provides technical assistance and publications on the transport and loading of hazardous substances. DOT, together with the National Fire Protection Association, sponsors a five-day course on handling hazardous material transportation-related emergencies. The course covers response techniques, hazards, coordination, and operating procedures.

U.S. Coast Guard

Although a Branch of the Armed Forces, the U.S. Coast Guard is under the jurisdiction of the Department of Transportation. The U.S. Coast Guard provides information and technical assistance in port safety. In environmental emergencies, the U.S. Coast Guard may provide a strike team and public information assistance. For persons involved in oil spill response, they offer a training course which includes spill planning and response and personnel safety.

Environmental Protection Agency (EPA)

The EPA provides technical assistance, publications, and analytical laboratory support. The Office of Emergency and Remedial Response is a major resource within EPA for technical assistance, training, and environmental protection support during hazardous substance incidents. The program areas of solid waste, water supply, and air pollution all have extensive technical experience and current research that directly relates to hazardous waste occupational safety and health. EPA, like OSHA, has regional offices which can

22

provide many of these services, or at least refer inquiries
to the appropriate office within the agency.

National Institute for Environmental Health Sciences (NIEHS): National Toxicology Program

NIEHS provides toxicological data, publications,
technical assistance, and research programs to study the
biological effects of potentially toxic substances found in
the environment. There is an emphasis on health effects
from chronic low-level exposures.

Food and Drug Administration (FDA)

FDA provides technical assistance, analytic resources,
and publications including the Pesticide Reference
Standards. The Food and Drug Administration is responsible
for the enforcement of EPA-established pesticide tolerances
through routine sampling and analysis of both the domestic
and imported food supply. This surveillance system is
designed to allow for prompt removal of products in which
the level of pesticide is in violation of the established
tolerance and to impose regulatory sanctions that will deter
repeated violations. Based on the results of FDA sampling
programs, it is evident that the vast majority of the
produce samples tested by the Agency comply with the law.
FDA's Bureau of Radiological Health is responsible for
research and technical assistance on methodologies for
controlling radiation exposures and on health effects from
these exposures. FDA, in cooperation with CDC, is
responsible for public health problems that are created by
hazardous substances in the food chain.

Library of Congress - National Referral Service

This service refers users to the appropriate
organization or program and publishes directories of
information sources in various subject areas including
physical sources.

American Conference of Government Industrial Hygienists (ACGIH)

ACGIH is not a Federal agency, but its membership is composed of industrial hygienists employed in government. ACGIH publishes annually a table of recommended threshold limits. This can often be obtained from NIOSH or OSHA. A series of Hygienic Guides on many commonly used industrial substances with references for additional information is also available.

DATA BASES AND HARD COPY REFERENCES

The data bases and reference volumes cited in this section are an abbreviated listing of the sources prepared for a forthcoming NIOSH manual on occupational safety and health at hazardous waste sites. This list includes only a few of the available Federal publications and data bases, and it excludes sources from the private sector in order to remain within the purview of this chapter.

Reference Books and Publications

Chemical Hazardous Response Information System (CHRIS). Four Volume Set. For sale by the Superintendent of Documents, Washington, D.C.

Hazardous Materials Spill Monitoring Safety Handbook and Chemical Hazard Guide. EPA 600/4-79-008a. EMSL/ORD, U.S. EPA, Las Vegas, Nevada, 1979.

Lewis, R. J. and Tatken, R. L., Registry of Toxic Effects of Chemical Substances, Volumes 1 and 2, Public Health Service, Centers for Disease Control, NIOSH, Cincinnati, Ohio, 1980.

Mackison, F., Stricoff, R. S., and Partridge, L. J., NIOSH/OSHA Pocket Guide to Chemical Hazards. U.S. Department of Health and Human Services, Public Health Service, Centers for Disease Control, January 1981. DHHS Publication Number 78-210.

Student, P. J. Emergency Handling of Hazardous Materials. The Bureau of Explosives, American Association of Railroads, Washington, D.C., 1980.

Hazardous Waste Sites and Hazardous Substance Emergencies.
Department of Health and Human Services, Public Health
Service, Centers for Disease Control, National Institute for
Occupational Safety and Health (NIOSH), Publication 83-100,
December 1982.

Hazardous Materials Incident Response Operations Training
Manual. U.S. Environmental Protection Agency, Office of
Emergency and Remedial Response, Hazardous Response Support
Division, June 1982.

Superfund: What It Is, How It Works. U.S. Environmental
Protection Agency Publication Number HW-1, April 1982.

Key, M., et al., Eds. Occupational Diseases - A Guide to
Their Recognition. U.S. Department of Health, Education and
Welfare, Public Health Service, Centers for Disease Control,
National Institute for Occupational Safety and Health,
Publication Number 77-181.

Criteria for a Recommended Standard for Hot Environments.
U.S. Department of Health, Education, and Welfare, Public
Health Service, Centers for Disease Control, National
Institute for Occupational Safety and Health Publication
Number 72-10269, 1972.

Eller, Peter M., Ed. NIOSH Manual of Analytical Methods
Third Edition. U.S. Department of Health and Human
Services, Public Health Service, Centers for Disease
Control, National Institute for Occupational Safety and
Health, Publication No. 84-100, Feb. 1984..

Data Bases

     Federal data bases exist in various forms: hard-copy,
microfiche, and electronic. Since Federal programs, like
the rest of society, are in a transitional phase of moving
from printed to electronic communication, the focus will be
on electronic systems. These are on-line interactive
computer systems maintained by various agencies. Some
systems can be accessed by requests to the responsible
agency or to other agencies, depending upon their
information retrieval systems. All are available through
one or more of the operational networks of chemical data
bases equipped with computer programs that permit
interactive searching and retrieval from these individual
bases. Commercially operated networks include ORBIT

(operated by System Development Corporation) and DIALOG
(operated by Lockheed Missiles and Space Corporation).
Access is often available through public libraries and
libraries within academic institutions on a fee basis. An
additional list of data bases is provided in Chapter 4,
Appendix I.

NIH-EPA Chemical Information System (CIS)

The CIS is a collection of computerized data storage
and retrieval modules for chemical information. The
components most relevant to occupational safety and health
include the Oil and Hazardous Materials/Technical Assistance
Data Base (OHMTADS), RTECS (which will be described below),
and Critical Toxicology of Commercial Products (CTCP) [6].

EPA - Solid Waste Information Retrieval System (SWIRS)

Operated by the Office of Solid Waste Management
Programs, SWIRS organizes all the published information
concerning current research and technological developments
in the solid waste management field worldwide.

NIOSH - The Registry of Toxic Effects of Chemical Substances (RTECS)

Maintained by the Division of Standards Development and
Technology Transfer (DSDTT), RTECS is an automated search
and retrieval system which is updated quarterly. RTECS is
also available in hard-copy and on microfiche. Currently,
there is information on approximately 44,000 compounds. The
information available on each compound includes toxicity
measurement, literature references, and specific health
effects.

NIOSH - Document Information Directory System (DIDS)

Also accessed through DSDTT, DIDS is an automated listing of all NIOSH publications, health hazard evaluations and internal NIOSH reports.

NIOSH - Current Research File (CRF)

CRF is a listing of on-going occupational safety and health research funded in the most recent three year period. This is also accessed through DSDTT.

Department of Commerce (DOC) - National Technical Information Service (NTIS)

NTIS is a cross-disciplinary file of citations to federally funded research and development reports. It includes analyses prepared by Federal agencies, their contractors or grantees and nonclassified Department of Defense reports.

National Library of Medicine (NLM) - MEDLINE

MEDLINE is the "on-line" portion of MEDLARS which serves as a guide to worldwide medical literature. It does not include abstracts. It corresponds to the printed Index Medicus.

NLM - The Toxicology Information Conversation On Line (TOXLINE)

TOXLINE accesses the current literature on human and animal studies on the toxicity of substances.

Emergency Hot Lines

The following programs are for emergency situations, and should never be accessed for routine information gathering.

U.S. Coast Guard - National Response Center (NRC)
1-800-424-8802

The U.S. Coast Guard mans the NRC which was established in response to the National Contingency Plan (40 CFR 1510). The NRC acts to: (1) receive reports required by law; (2) insures rapid response by notifying the appropriate government authorities when a pollution incident occurs, and (3) provides advice and assistance via the Chemical Hazard Response Information System (CHRIS), the Hazard Assessment Computer System (HACS), the Hazardous Materials Emergency Response System (HMER), and the Office of Hazardous Materials Technical Assistance Data System (OHMTADS).

Since the NRC functions as a focal point for notification and coordination of offices within the government and the private sector, it can be used for almost all hazardous material accidents. Potential users are strongly urged to become familiar with those situations where "immediate" reporting to the NRC is mandatory.

To obtain the most rapid response, the U.S. Coast Guard recommends that the caller provide as much of the following information as possible (this recommendation is applicable to any Federal, private, or state "hot-line" or emergency numbers) [7].

  * name;
  * telephone number where the caller can be reached;
  * location of emergency and/or body of water affected;
  * type of vehicle, facility, or container involved and amount spilled;
  * rail car or truck number and chemical identification number, U.N. code, etc.;
  * carrier name;
  * shipper or manufacturer;
  * consignee;
  * local weather conditions;
  * any response/cleanup actions presently in progress or planned; and
  * extent of injuries, if any.

Established by the Chemical Manufacturers' Association
(CMA), CHEMTREC serves as a clearinghouse for chemical
transportation emergencies. It serves as a liaison between
the user, shipper, and manufacturer, and as a contact point
with other industry groups such as the Chlorine Institute
and the Fertilizer Institute.

Although a private institution, CHEMTREC works closely
with the National Response Center (NRC) and with the
Department of Transportation. It should be emphasized that
requests to CHEMTREC do not fulfill Federal reporting
requirements.

ACCESSING FEDERAL PROGRAMS

Government publications and resources in any subject
area present problems for their actual and potential
audiences, due to their great number, their different
distribution channels, and the number of organizations and
programs producing them. In subject areas like hazardous
waste, these potential problems are exacerbated by the
dispersion of responsible agencies throughout the federal
bureaucracy, the multidisciplinary nature of the subject,
and rapid growth resulting from public concern and current
legislation. Although these factors may complicate access,
they are also indicators of the magnitude and diversity of
the available resources.

Once potential users identify these and other potential
problems, strategies for accessing federal programs can be
developed. Although each user will develop his or her own
unique approach, the most effective strategies will be based
upon knowledge, planning and interpersonal skills.

Forethought will result in better service, less
frustration, and less time expended. Planning will enable
the user to obtain some familiarity with the legislation and
programs related to the inquiry. It will also allow enough
lead-time for proper definition of the inquiry and contacts
with local resource people. Potential users are strongly
urged to consult with librarians and information specialists
at public, academic, and industrial libraries. A recent
article about on-line sources for Toxicology and Safety Data
clearly indicates the changing nature of Federal government
information systems and the need for information specialists

[8]. These individuals are not only trained professionals; they are also "users" themselves with established methods of access. They can save both the potential and experienced user considerable time and frustration. Frequently, they can also provide sample output from automated information systems and refer users to specific individuals.

Planning should also include definition of the kind of information or services needed. Given the volume and diversity of the resources available, the user will be best served when the objective of the inquiry is clearly identified. The following questions should be considered before making any inquiry:

1. Is the most current information needed or is older information applicable?
2. Is a search of the national or worldwide literature needed? Can the search be limited to specific books or journals?
3. Is the required information discipline-oriented, multidisciplinary, or problem-oriented?
4. Which of the following will be needed: a bibliography, an annotated bibliography, or actual documents?
5. What is the depth of information needed?
6. When is the information needed?

Clearly-defined inquiries result in a better level of service. Even when made to the wrong program, they aid Federal employees in referring users to more relevant or supplemental programs. The well-defined inquiry probably has a strong beneficial impact on the interpersonal level, since Federal employees receive global or poorly-defined requests everyday, e.g., "What businesses and things does OSHA regulate?" or "What kind of training courses are available?"

These interpersonal contacts are important and should be cultivated. Through these referrals, potential users can develop a network of contacts within a number of federal agencies. As these contacts are developed, access becomes more efficient. The Federal employee may become more familiar with the user's needs, while the user knows which agency has what, and how long it takes to obtain it.

Persistence in follow-up is as important in acquiring Federal program assistance as it is in the sales business. Program changes and staff turnover necessitate a relentless pursuing and building of these communication networks. The user's skill and knowledge can be used and increased only if contact is maintained with individuals within the Federal program. It is only through these contacts and dialogues that users obtain the most comprehensive and current level of assistance available.

One common characteristic shared by all Federal programs is change. All are subject to constantly changing personnel, available resources, program emphasis, and administrative organization. These can range from minor modifications of the administrative organization to major restructuring of the program emphasis. Since even these minor modifications can result in personnel transfers and new telephone listings, users must monitor the program fairly frequently. A listing of Federal programs, addresses, and telephone numbers are provided as a good starting point (see Table I). Professional journals, newsletters and the Federal Register all report changing authorities, funding levels, and key staff changes. Abstracting services can be purchased to monitor Congressional action affecting budgetary restraints, program modifications, and new regulations.

Keeping up to date with any specific Federal program or group of programs requires obtaining pertinent information from the voluminous amount available. This is particularly true with changes in the Federal administration. The outgoing administration attempts to push through legislation and programs in its final days (e.g., Superfund) and the new administration makes changes in key staff and administrative procedures in order to institute its own agenda. During this period of the "changing of the guard", there is considerable discussion and debate all of which contribute a huge amount of information that must be sorted through to track any actual program changes.

The task of obtaining pertinent information from this largess is even more formidable with high level public interest programs like Superfund. Their potential social, economic, and environmental impact can mobilize many special interest groups which further contribute to the available information. These groups can also affect rapid changes within Federal programs through their support and lobbying activities.

Rapid changes in the program implementation of both Superfund and the Resource Conservation and Recovery Act of 1976 (RCRA) have required outside users to keep close watch in order to know what services are available, the limitations on Federal programs, and the new requirements established by regulatory amendments. This has been further complicated by variable levels of state legislative activity. As the states assume more responsibilities under both acts, legislative activities associated with RCRA and Superfund will continue to be substantial [9].

Table I. A Partial List of Governmental Organizations Addresses,
and Telephone Numbers that can Provide Additional
Hazardous Waste Health and Safety Information

Centers for Disease Control (CDC)
Center for Environmental Health
Superfund Implementation Group
4770 Buford Highway
Chamblee, GA 30341
(404) 452-4100

Department of Energy (DOE)
Office of Scientific and
  Technical Information
P.O. Box 62
Oak Ridge, Tennessee 37831
(615) 576-1301

Department of Transportation (DOT)
400 Seventh Street, SW
Washington, D.C.
(202) 426-2075

Environmental Protection Agency
  (EPA) Office of Emergency and
  Remedial Response
401 M Street SW
Washington, D.C.
(202) 382-4770

Federal Emergency Management
  Agency (FEMA)
500 C Street SW
Washington, D.C. 20472
(202) 287-3904

Food and Drug Administration (FDA)
5600 Fishers Lane
Rockville, MD 20857
(301) 443-3170

Library of Congress
National Referral Center
Washington, D.C. 20234
(202) 287-5670

National Bureau of Standards
Washington, D.C. 20234
(301) 921-2318

National Institute for
Environmental Health Science
(NIEHS)-National Toxicology
  Program
P.O. Box 12233
  Research Triangle Park
  North Carolina 27709
(919) 629-3991

National Oceanic and
  Atmosphere Admin. (NOAA)
U.S. Dept. of Commerce
6010 Exec. Blvd.
Rockville, MD 20852
(301) 443-8963

National Institute for
  Occupational Safety and
  Health (NIOSH)
1600 Clifton Rd. NE
Atlanta, GA 30333
(404) 329-3771

Occupational Safety and
  Health Admin. (OSHA)
Department of Labor
200 Constitution Ave. NW
Washington, D.C. 20210
(202) 523-7505

U.S. Coast Guard
Commandant (G-WER-12)
Department of Transportation
Washington, D.C. 20593
(202) 426-9568

U.S. Geological Survey (USGS)
National Center
12201 Sunrise Valley Dr.
Reston, VA 22092
(703) 860-6167

Although most government programs are subject to constant modification, programs currently involved with hazardous waste are probably subject to a much higher rate of change. The current level of public concern about hazardous waste has made it highly visible and politically sensitive.

These factors and the potential economic implications of hazardous waste management and remedial actions will result in further Federal program modifications for some time. Even though this rapid rate of change will be a source of frustration for potential users, those tenacious enough to pursue their inquiries will be rewarded with the resources derived from these program changes.

REFERENCES

1. 5 U.S.C. 552.
2. Bennett, G. F., et al. Hazardous Materials Spills Handbook (New York: McGraw-Hill Book Co., 1982), p. 42.
3. "General Industry," U.S. Department of Labor, Occupational Safety and Health Administration, OSHA 2206, U.S. Government Printing Office, (Revised June 1981), p.6.
4. Martin, William F., J. M. Melius, and C. A. Cottrill. "Management of Hazardous Wastes and Environmental Emergencies," paper presented at the National Conference and Exhibition on Hazardous Wastes and Environmental Emergencies, Houston, TX: March 12-14, 1984.
5. Wallace, Lynn P., W. F. Martin, and C. A. Cottrill. "An Overview of NIOSH's Preparation of Hazardous Wastes Worker Health and Safety Guidelines," paper presented at 4th National Conference on Management of Uncontrolled Hazardous Waste Sites, Washington, D.C., October 31-November 2, 1983.
6. Heller, S.N., and G.W. Milne. "Use of the NIH-EPA Chemical Information System in Support of the Toxic Substances Control Act," in Monitoring Toxic Substances, D. Schuetza, Ed. (Washington, D.C.: American Chemical Society, 1979) pp. 255-286.
7. "National Response Center" U.S. Department of Transportation, United States Coast Guard.
8. Page, N. P., and H. H. Kissman. One-Line Sources for Toxicology and Safety Data, presented at National Conference and Exhibition on Hazardous Wastes and Environmental Emergencies, Houston TX: March 12-14, 1984.
9. Hazardous Waste Management: A Survey of State Legislation 1982 (Denver, CO: National Conference of State Legislatures, 1982).

3

OCCUPATIONAL HEALTH AND SAFETY PROGRAMS FOR HAZARDOUS WASTE WORKERS

David L. Dahlstrom
Corporate Director of Safety and Health
Ecology and Environment, Inc.
195 Sugg Road
Buffalo, New York 14225

The rapid accumulation of hazardous wastes poses one of the most complex and expensive environmental control and cleanup tasks in history [1]. For this reason, numerous environmental statutes were enacted in the decade of the 1970's which have had a far-reaching effect upon industry, society, and the environment [2-10]. While these statutes and regulations have prescribed the means and methods to be followed to prevent significant deterioration of the overall environment in which we live, as well as to insure the health and well-being of both the public at large and the employee within the conventional workplace, in many respects the statutes have neglected to prescribe the means and methods necessary to properly protect the employee who must investigate, handle, dispose, and control hazardous wastes [11]. To carry out these tasks, health and safety professionals in the newly emerging hazardous waste industry are in the process of adapting the administrative, engineering, and personal protective controls of the conventional workplace to a unique setting. This chapter presents the current state of this technology as it is applied toward the ultimate goal of insuring the continued good health and well-being of those who work at hazardous waste sites. In particular, this chapter demonstrates the need for a concerted approach to the problem involving the multiple disciplines of industrial hygiene, industrial safety, toxicology, engineering, and medicine.

Administrative Policy and Goals of a Successful Program

     The development and implementation of a comprehensive
health and safety program for the hazardous waste employee
is a complex and interactive effort for which an
industry-specific set of codes are nonexistent, and few, if
any, guidelines exist. As is the case for most companies,
the degree of development of such a program depends upon the
size of the facility (operating site), the number of
employees involved, the types of operations being conducted,
the variety of materials and potential hazards encountered
at the work site, the type of business (consultant, waste
transport, treatment, or disposal), and most importantly--
management's overall philosophy towards health and safety.
Irrespective of these variables, however, most successful
programs have a policy on occupational safety and health
which entails certain key requirements. These requirements
are listed below.

1. A policy which is explained and made available to
    all employees in order to insure a thorough
    understanding of program goals and individual
    responsibilities.
2. A clear definition of program objectives and a
    schedule for achievement.
3. An overall commitment to support a program which
    acknowledges to the worker management's
    responsibilities to the program.
4. A mechanism which will provide for mutual
    representation from all functional levels within
    the organization in the setting of priorities and
    the implementation of program objectives.
5. A clear definition of line and staff
    responsibilities and their reporting
    relationships. This is often best accomplished
    through the use of a functional organizational
    chart.
6. A means of periodically reviewing progress and
    accomplishments over the course of the program.

     Just as company programs vary in their size and scope,
so too do the overall goals of the program vary. Basically,
however, it is important that the goals set by the company
reflect the priorities of that company with respect to
employee health and safety. Developing the goals and

objectives for a health and safety program should not be a unilateral effort. The health and safety professional should solicit the assistance and input of colleagues involved in industrial hygiene, toxicology, engineering, biostatistics, epidemiology, medicine, and especially from the management team most affected. Further, these goals should serve as the basis or foundation for specific company policies and operating procedures, and should provide the philosophical framework for setting more specific objectives. It is by attaining these goals and objectives that the effectiveness of the program can be evaluated over time.

The basic goals listed below, while not all inclusive, will insure the continued good health and well-being of all employees who work routinely with hazardous materials.

1. Maintaining a safe and healthful work environment by placing all personnel in jobs according to individual physiological and psychological makeup, experience, and educational background.

2. Insuring compliance with all legal requirements mandated by the various federal, state, and local environmental and occupational safety and health regulations

3. Educating the employee in the proper application of company health and safety operating procedures and the use of associated equipment to insure a thorough understanding of the whys and wherefores of the job.

4. Limiting company and personal liabilities associated with the hazardous waste industry due most commonly to negligence. This can be accomplished through the close adherence to the concepts and requirements of such environmental and occupational laws as the Toxic Substances Control Act (TSCA), the Resource Conservation and Recovery Act (RCRA), the Occupational Safety and Health Act (OSHA), the Comprehensive Environmental Response, Compensation and Liabilities Act (CERCLA or "Superfund"); and the ever increasing, precedent setting cases known as "Toxic Torts" [3,8,9,10].

It is important that these goals correctly address the priorities and principles of the company, that they are clearly stated, and that they are updated as necessary.

The first major task in developing and implementing a comprehensive program is providing justification for its existence. All too frequently, the problem of absorbing the costs associated with a comprehensive health and safety program has proven to be a major obstacle to its successful development and implementation. This is especially true in the case of smaller companies, which generally turn to a consulting firm specializing in health and safety to provide these services on an as-needed basis [12,13]. In any case, the problem of justification can be approached from the aspects of economics and the underlying responsibility of the company to its employees and the surrounding community [14].

It is the responsibility of the health and safety professional to demonstrate to a management concerned with the "bottom line" that there is a need to balance the amount of resources invested and the return on that investment with the potential and actual costs associated with what we will call "negative" economics. Here, we are referring to those costs associated with the legal liabilities pertaining to the toxic tort or personal damage suits. The settlement awards and court costs in these instances can be quite enormous. It is important to recognize further the ancillary costs associated with personal injury or regulatory compliance suits. These are the costs incurred due to the eventual increase in insurance premiums, property losses (private or internal), and operational stoppages. There will be costs associated with labor problems which may result if the work site is perceived as being unsafe or unhealthy, and the costs associated with poor productivity due to low employee morale. Also included will be the real and administrative costs of paying wages to injured workers who are not covered by insurance but are part of a collective bargaining agreement, as well as the costs attributable to regulating penalties and eventual OSHA surveillance [15]. Considering all these factors, it is far less costly for a company to invest in a comprehensive health and safety program that is preventative in nature than to wait and hope an accident does not occur.

Regarding the responsibility of the company to its employees and the surrounding community, various environmental and occupational statutes require an employer to provide its employees with a safe and healthful workplace, as well as insuring that the surrounding environment is not adversely affected as a result of poor facility operations. Most companies within the hazardous waste industry desire to maintain a good community image and are sincerely interested in the well-being of their employees. Most company managers readily recognize the potential impact which adverse publicity can have on the future of a licensed waste treatment/handling facility or cleanup company and to employee efficiency and productivity. To avoid negative publicity and to maintain good employee relations, management should strive to communicate to its employees and the public the positive actions being taken by the company to provide acceptable levels of health and safety protection [16]. A sound health and safety program is just such an action and can be used to demonstrate the commitment of the company to protecting employee and public health and the environment.

Therefore, a total commitment by management in terms of providing support and dollars in the development and implementation of a comprehensive health and safety program is well justified. This function of management will serve not only as a preventative measure in the protection of its workers and the surrounding community, but also in the protection of the good name and assets of the company.

Encouraging Worker Commitment to Health and Safety

Just as a total commitment by management is integral to the ultimate success of a comprehensive health and safety program, so too is the commitment of the work force. This commitment by the work force is dependent upon the manner in which policies are applied, and hence perceived, by the employee. To assist in creating a positive perception, it is essential to encourage employee involvement in achieving program goals and objectives [17,18]. This can be done by incorporating into the program the elements listed below. The incorporation of these elements will provide a solid foundation upon which a successful and meaningful program can be built.

Creating a work environment that is as safe, healthful, and as free from recognizable hazards as possible. Every effort must be made to provide the employee with proper and adequate equipment with which to do the job. These efforts should be coupled with the installation of various engineering controls (e.g., positive ventilation, or dust suppression) to minimize potential exposures. It is difficult to motivate employees to adopt safe work practices or to use protective equipment if management fails to provide an adequate work environment or fails to institute controls to protect employees from exposure to hazardous chemicals or operations.

Insuring that there exists within the program a means for clear and open communications. This should allow employees and management alike to communicate not only problems and recommendations, but also any changes in program policies and procedures. This can be best accomplished by including employee representatives chosen by their co-workers, on the health and safety committee of the corporation. If suggestion boxes are used for purposes of communication, management must respond in a sincere and meaningful fashion to reinforce the credibility of the program and management's commitment to it; yet, it must be stressed that communications must be two way.

Establishing both staff and line responsibilities which will facilitate the fulfillment of the goals of the program. Line management must be given the direct responsibility of insuring employee health and safety and preventing needless property loss. In order to fulfill their responsibilities in the compliance with the various program policies and procedures. the health and safety professional must be given the staff responsibility of assisting line management and the employee.

The goals and objectives of the overall program should revolve around the concept of risk evaluation. Management has the ultimate responsibility of maintaining an accurate assessment of the associated health and safety risks within each work site [14]. Management must develop the means of dealing with these defined risks in a reasonably scientific and objective manner. This can be done through the determination of "acceptable risk", which entails applying technically acceptable means of quantitating and qualitating the chemical and physical hazards of a site and the subsequent application of the various occupational regulations and associated standards [19]. Through the application of sound industrial hygiene techniques in monitoring both the work site environment and personnel in a consistent and conscientious manner, the proper risk evaluation can be made.

Informing the work force of the recognized risks within the work place. Many states and localities now require compliance with the requirements of established worker "right-to-know" laws. Informing employees of the risks associated with their job and demonstrating that these risks are being minimized through proper controls will encourage employee cooperation toward the accomplishment of program goals and objectives.

FIVE ESSENTIAL COMPONENTS OF THE HEALTH AND SAFETY PROGRAM

The specific components of a health and safety program are actually the bricks and mortar which solidify and structure it. They provide the means by which the employee can be assured of continued good health and a safe environment in which to work through the incorporation of administrative and engineering controls, and the use of personal protective equipment. These controls instituted

from the aspect of prevention, serve to minimize the
potential for accidents and overt exposures while maximizing
the professionalism and proficiency of the employees. Of
course, the degree of conscientiousness in which these
components are applied will determine the structural
integrity (and therefore the success) of the program.

Within the hazardous waste industry, the following
program components have served to insure the quality of
company health and safety programs:

1. A comprehensive employee health surveillance
    system;
2. A safety program consisting of specified company
    guidelines and standard operating procedures, and
    the assignment of both company and site-specific
    safety coordinators whose responsibility is to
    insure compliance with program requirements;
3. An industrial hygiene program with conventional
    monitoring and evaluation techniques adapted to
    the unique setting of the hazardous waste site;
4. A company health and safety advisory committee
    whose purpose is to assist the company in
    developing, maintaining and periodically
    evaluating a state-of-the-art health and safety
    program based on current technological advances;
5. An in-depth, hands-on training program which
    includes periodic refresher training.

The remainder of this chapter discusses each of these
components, except for number 5, "Training" which is the
topic of Chapter 12.

HEALTH SURVEILLANCE SYSTEM

The health surveillance system should prescribe
specific fitness criteria in conjunction with periodic
medical evaluations to insure that only medically sound
individuals participate in field operations involving
hazardous materials and that the health status of these
individuals is maintained. Therefore, the purpose of this
system is to detect any changes in the health status of
individual workers or employee groups which might be related
to the nature of the job performed and the substances with
which the employee comes in contact. In consideration of
the nature and variety of chemicals (and mixtures thereof)
related to hazardous wastes and the frequency for potential
exposure to these chemicals, it is extremely important to

provide a means by which the health of personnel who work at hazardous waste sites can be periodically assessed.

To properly monitor the health status of each employee, the specific items of data needed must first be identified. Then, it must be determined how this information is to be most effectively collected, collated, analyzed, and used. It is essential that these steps be completed during the preplanning stages of surveillance system implementation so as to prevent the common mistake of collecting too much information (a scatter-gun approach), some or most of which eventually proves to be unnecessary or inapplicable. Such a mistake can be costly in terms of both money as well as time and materials spent. Only essential pieces of data that will prove most beneficial and revealing respective of individual changes in health status should be collected. Essential information to be collected is discussed below.

Complete baseline data on the health status of the individual at the time of employment. This should include information on: 1) family and individual medical history; 2) prior work history, including the types of chemicals the individual has worked with; 3) any known individual abnormalities; personal habits, such as the use of alcohol, tobacco, or drugs; 4) prior surgeries, hospitalizations, and immunizations; 5) reproductive history; and 6) any abnormalities found during the initial preemployment physical examination and laboratory screening.

Nature of the work to be performed. This data should include types of activities to be performed, the types of chemicals to be encountered, exposure data as collected through site and personal monitoring, and accident data.

Individual identification data. This should include date of employment, name and social security number, date of birth, race and sex, and current address and phone number.

Follow-up data. This will include data collected during subsequent periodic (i.e., annual) physical examination and laboratory screenings.

The orderly collection of these data on a form designed with a coding scheme will permit the information to be easily collated and analyzed on a computer, thereby easing the burden of OSHA required recordkeeping. In addition, it will insure the confidentiality of medical records and provide for rapid access to records in case of emergency.

It cannot be overemphasized that the manner in which the data is collected and analyzed be well thought out and designed so as to insure its complete usefulness. It should also be recognized that the data must be complete. The same data must be collected for each member of the work force. If this is done appropriately, the relationship to work performed or exposures experienced and health status of the

42

individual and the group can be easily and effectively correlated.

At this point, it should be clear that the development of an employee health surveillance system requires significant preliminary work. The necessary preliminary work includes the design of data collection forms, the development of a system for coding and collating data for computer input, the development of the software necessary to generate the necessary reports, and the design of procedures to monitor data processing and report flow in order to meet defined schedules. The system also requires that a specific individual be designated as responsible for coordinating all of these activities. This individual may either be a company employee, such as the corporate medical director or health and safety director, or a medical consultant from outside the company. The important factor is that a responsible person be designated to insure the smooth operation and the quality of the system so as to provide the employee with the maximum level of protection.

Once the requirements of data collection, collation, analysis, and retention have been determined and a responsible organizational party identified, the components of the system can be easily defined. These components should include:

1. Each worker should be given an initial medical examination and their medical history assessed prior to the performance of hazardous waste site activities. This will establish each worker's medical baseline and overall health status and insure that each individual is capable of undertaking the rigors of field operations. It will also aid in determining which job is best for that worker.

2. Each worker should undergo subsequent medical examinations on a periodic basis (usually annually). The examination should be geared toward the worker's job classification and the chemicals which the worker confronts. The parameters of this examination should be consistent with those of the initial examination in order to insure consistency.

3. An exit examination should be required of each individual either leaving the company or transferring to another job within the company not associated with hazardous materials. The results of this examination can serve to document the health status of the individual at that time.

4. A contingency plan should be instituted which would require the monthly reporting of exposures

or injuries. It should also specify the design
and performance of specific post-exposure
examination protocols based on the types of
chemicals to which the employee is exposed. The
plan must include an emergency medical plan, which
would include provisions for informing the
attending physician with pertinent information
regarding the affected individual and
toxicological information specific to the
chemical(s); this will facilitate proper diagnosis
and treatment. There should be in place an
emergency analytical system which will quickly
analyze unknown chemicals to which an individual
has been exposed so as to provide the physician
and the toxicologist with the identity of these
chemicals. The plan should also specify: 1) a
means for removing any women from the infield work
force who become pregnant so as to provide for
fetal protection; 2) the provision of specific
tests for male and female employees to insure
their reproductive health; and 3) the removal from
work site responsibilities of any worker who shows
a significant abnormality in their medical
profile, at least until it has been determined
that he or she has completely recovered and is in
good enough health to resume work responsibilities
[20,21,22].

The implementation of such a health surveillance
program is neither a trivial nor an impossible undertaking.
It does require considerable attention and coordination in
order to be successful, but the benefits derived from a
system that has been designed to be preventative in nature
are innumerable in terms of employee health, welfare, and
productivity.

SAFETY PROGRAMS

The purpose of a safety program is to protect the
employee from the hazards, especially physical hazards,
during hazardous waste work. Currently, no specific
guidelines or uniform code specific to the hazardous waste
industry exist for the development and implementation of a
safety program. Basically, a safety program should
complement and support the industrial hygiene and health
surveillance systems and it must be an integral part of
every aspect of site operations. Most successful safety

programs, despite variations in organizational structure and technique of application, make safety a major priority with respect to company policy and action [23]. This attitude of management, if demonstrated in a consistent and sincere manner, lends credibility to the program, encourages employee cooperation and support, and most importantly, minimizes the frequency and severity of accidents.

The critical elements of a successful safety program are:

1. Significant employee involvement in the development and implementation of operational procedures;

2. Openly demonstrated and consistently applied management support and leadership;

3. Assignment of responsibilities to specific persons within the corporate, divisional, and facility structures whose role it is to insure compliance and employee understanding of company policies and procedures;

4. Development and implementation of standard operating procedures covering all aspects of the work performed;

5. A personal protective equipment plan which defines the decision-logic necessary to insure that proper respiratory and dermal equipment is provided and also provides for its proper use and maintenance;

6. A safety communications system which provides for the dissemination of information to all employees regarding hazard identification, policy and procedural changes, and lessons learned;

7. A comprehensive training program which provides for periodic classroom and hands-on training of the employee in the proper use of equipment, operational techniques, and company policy;

8. An accident record and investigation program that not only will satisfy all legal requirements, but also will prevent a similar reoccurrence;

9. An audit program to insure consistent application of operational procedures and evaluate their effectiveness; and

10. The maintenance of safe working conditions.

Many of these elements are self-explanatory and require no further discussion. The remaining elements are discussed in greater detail in subsequent chapters. It should be kept in mind, however, that the role played by the safety professional in employee training often determines the effectiveness of the other elements and deserves further consideration in this chapter.

The importance of assuring that the safety program maintains a high level of visibility has already been alluded to. It therefore becomes the role of the safety coordinator to insure this visibility at all levels within the organizational structure. All individuals involved in safety coordination, from the health and safety director of the company to the site safety officer, have the responsibility of insuring the total compliance by the employee with all of requirements, policies, and procedures of the safety program. Safety coordinators must develop, implement, and evaluate the effectiveness of site-specific safety plans, operational procedures, equipment usage, and employee performance within the work setting. In cooperation with the industrial hygienist and the engineer, the safety coordinator must ensure the application of proper site and personal monitoring techniques so as to provide the employee with pertinent information regarding site hazards and the proper selection of appropriate protective equipment. In this way accidents can be prevented, thereby protecting the health and welfare of the employee and the surrounding public. In the event of an accident, the safety coordinator must investigate its occurrence and prescribe preventative measures to preclude its reoccurrence, and should maintain records of all accidents on-site to determine accident trends. Such records are required by OSHA and, no doubt, company management. Further, the safety coordinator must strive to instill among employees a high degree of safety consciousness through insuring employee participation, communication, and education.

To assure that all employees are prepared to safely participate in all facets of waste site operations, a series of specific and specialized training programs must be established as an integral part of the overall health and safety program. These programs should include training the employee in a classroom setting as well as in a practical setting to allow hands-on experience. Periodic refresher courses are also important.

Prior to performing work on-site, each employee should undergo training in a classroom setting. This training should provide the new employee with information regarding: 1) company policies and procedures; 2) health and safety requirements; 3) basic toxicology and chemistry of chemicals commonly confronted on waste sites; 4) the selection, use, and limitations of the various respirators and dermal protective equipment; 5) techniques used in the decontamination of personnel and equipment; 6) sampling techniques; 7) legal requirements under "Superfund", OSHA and RCRA; 8) heavy equipment operation (if applicable); 9) the health effects of heat and cold; and 10) emergency procedures, including multimedia first aid and cardiopulmonary resuscitation.

Upon completion of the in-depth classroom training, each employee should be given the opportunity to gain actual hands-on experience in a field setting, where the worker learns to put classroom concepts into practice. This experience will allow each employee to gain confidence in themselves as well as the equipment and procedures upon which they ultimately must rely. Moreover, this type of training builds on the "team" concept which is so important during work on hazardous waste sites.

Refresher training should be provided periodically. The purpose of this training is to keep the employee informed of new techniques, policies, and procedures as well as to improve upon skills previously learned. The frequency and setting of refresher training will vary depending upon the diversity of job settings and each employee's ability to assimilate the training provided.

The application of the elements presented earlier, in conjunction with consistent management support, will serve to maximize the employee's proficiency and professionalism in conducting work operations at hazardous waste sites.

INDUSTRIAL HYGIENE PROGRAM

During cleanup activities at hazardous waste sites, the primary objective is to minimize health hazards to site workers and the surrounding general public. To achieve this, a site safety plan providing specific standard operating procedures must be developed. An effective site safety plan should address three key issues: identification of substances on-site and the hazard they represent; evaluation of the risk associated with those substances; and control of their potential impact on site personnel. Identifying substances and hazard evaluation generally involves reviewing historical and monitoring data obtained during preliminary assessment of site conditions. These data may include information obtained from manifests or facility records, as well as the results of off-site air monitoring and off-site drainage or leachate samples. Data from the identification and evaluation process provide the basis for evaluating exposure risks and determining measures to control potential impacts of exposure. These include ambient and personal monitoring, the proper selection and use of a variety of personal protective equipment, medical surveillance, safety training, and contingency planning. Proper consideration of these safety issues prior to on-site activity will result in more effective site operations [24].

The words identification, evaluation, and control in

the preceding paragraph allude to the role of the industrial hygienist who, in conjunction with the safety coordinator and toxicologist, is responsible for assisting in protecting the worker and the general public from the hazards encountered at hazardous waste sites.

The health and safety of the worker and the public should be of primary concern in all phases of investigative and remedial activity at hazardous waste sites, from the most routine site survey involving air, water, soil, or waste sample collection to the most complex site excavation or waste treatment schemes. Therefore, the scope and sophistication of the investigation activity, plus the level of on-site effort, largely dictate the breadth of the industrial hygiene services necessary. Effective implementation requires application of realistic protocols for hazard recognition, evaluation, and control, tied closely to the risks associated with the hazard potential posed by the site. These risk evaluations can vary daily or more often as a project progresses: workers will move to new locations on-site which will result in changes in their proximity in relation to contaminated zones, and/or modification of work practices. As a result, the industrial hygiene and safety monitoring performed during investigative and remedial work at a hazardous waste site differs from conventional industrial hygiene surveys in several ways, including: 1) the varied scope of safety and health concerns involved in such an effort; 2) the need for real-time as well as time-integrated analytical data; 3) the dual focus (occupational and community) of the air monitoring program; and 4) the unique, multimedia, often unknown, and difficult to quantify composition of the chemical contaminants.

It therefore becomes the role of the industrial hygienist to provide continued input during site operations to insure operations are conducted safely. Early coordination during project planning is essential to integrate safety and industrial hygiene procedures into the operational aspects of the work plan. The industrial hygienist assists the safety coordinator in developing a safety plan that is specifically tailored to the level of effort and the hazards associated with the work. (This occurs only after an exhaustive data review of background information and relevant toxicological data has been conducted, and the conceptual operations plan that defines the scope of work has been clarified.) The criteria used in the selection of the appropriate scope of the industrial hygiene and safety field protocols prescribed in the safety plan include toxicity-related factors and exposure potential factors [25]. Each of these factors should be defined and considered in conjunction with each of the other factors so as to best characterize the site in a comprehensive manner.

Toxicity-Related Factors	Exposure Potential Factors
* chemical agents	* job function
* concentrations (background, episodic	* proximity to zones of contamination
* dose-response relationships	* accident/major release potential
* physiologic/synergistic consequences	* level of site activity
* TLVs, ceiling limits, STELs	* physical properties of the agents
* odor thresholds	* frequency of exposure
* percutaneous characteristics	* route of exposure
* carcinogenic/mutagenic/ teratogenic properties	* atmospheric dispersion characteristics

In addition to these risk assessment factors, the practical concerns of instrument limitations and sensitivity, sampling frequency and duration, and logistical implementation of the protocols will influence the overall focus and effectiveness of the industrial hygiene and safety program developed for each site.

As the investigation and eventual cleanup of the hazardous waste site progresses, the role of the industrial hygienist expands to address the various objectives of site operations. These objectives include: 1) the upgrading and downgrading of the levels of dermal and respiratory protection on the basis of the chemical contaminants encountered and their concentrations within the worker's breathing zone; 2) documentation of ambient air and emission episodes for recordkeeping and information planning purposes; 3) on-site sample characterization using portable gas chromatographs, photoionization, and infrared devices for the purposes of prescreening samples so as to reduce the analytical loads or to more accurately identify constituents on a real-time basis; 4) monitoring of on-site personnel both for potential chemical exposure and the effects of heat stress or fatigue; 5) the specification of engineering, administrative, or personal protective controls to mitigate any unacceptable hazards; and 6) recommending corrective actions and subsequently evaluating their effectiveness to prevent exposures to on-site or off-site locations beyond the predetermined action levels designed to protect worker or public health. Generally, separate instruments or monitoring procedures must be used to address each of these objectives because of the variable locations or time frames in which the data are needed.

In summary, the aspects of health surveillance, industrial hygiene, and safety must be integrated into a single program. The activities of each should serve to interrelate and support each other. Therefore, it is imperative to the success of the overall program that the roles and responsibilities of individuals in each of these areas be closely integrated.

STAFFING AND ORGANIZATION

Size and Qualifications of Health and Safety Staff

The size and organization of a health and safety staff of a company will depend upon the size of the company, the types and variety of hazards, and the amount of resources available for salaries and consultant services. Generally, the larger the company, the greater the need to develop in-house health and safety capabilities. Many companies find it more practical initially to hire a consulting service during the preliminary stages of program development. A consulting company may also be needed for additional support during the various stages of program implementation and for assistance with specific problems which may occur once the program becomes operational. Once the program is running smoothly, however, the use of consultants is generally relegated to those situations where their services simply augment rather than substitute for in-house capabilities.

As mentioned previously, the health and safety program staff is generally composed of trained individuals from the fields of medicine, industrial hygiene, toxicology, engineering, and safety. The number of individuals within an organization possessing these capabilities will depend upon the size and needs of the company. Due to the current demands for such trained and experienced individuals, their acquisition may be difficult, if not initially overwhelming. Many companies therefore attempt to train existing staff members in some of these areas through the use of short courses or through company programs for degree level studies.

At a minimum, the health and safety staff should include an occupational physician (full-time or consultant), a company safety and health director, and industrial hygienist, and a toxicologist (full-time or consultant). The role of the occupational physician, who may either be a full-time employee in the case of a large company, or more commonly, a doctor from a nearby hospital specializing in

50

occupational medicine, is to provide the necessary examination and emergency services as well as to assist in the continued development of a sound health surveillance program.

The health and safety program director is responsible for the overall coordination and operation of the complete program. This individual generally reports directly to the company vice-president in charge of environmental affairs. He or she must understand the goals and objectives of the company and be able to develop and implement the necessary programs to achieve them. It is especially important that the director be given the authority to implement programs and procedures, to acquire the necessary funding and to make expenditures to maintain the program, and to delegate responsibilities to other personnel through their supervisors. The director must be familiar with the procedures and materials regarding all workplace operations and possess the knowledge to assess them properly based upon the principles of occupational safety and health.

The director will usually be responsible for supervising a staff of industrial hygienists and safety professionals whose composition again is dependent upon the needs and the size of the company. These professionals obviously should be well trained and experienced, and must be thoroughly familiar with the types of operations being conducted on hazardous waste sites. Their roles within the company will be to fulfill the specific objectives of insuring worker health and safety at particular waste sites and to coordinate with and support the efforts of the health and safety director. Generally, larger companies will hire separate individuals to fill these positions, while small companies will commonly rely upon individuals who possess these dual capabilities.

Obviously, a certain number of support staff will be necessary to assist the health and safety staff in the day-to-day operation of the program. Primarily, clerical staff can fill this gap as well as assist in program communications and budgeting.

Budget

The health and safety director is usually made responsible for the preparation of the budget for these activities, receiving assistance from the accounting and clerical staffs. In order to develop a realistic budget which is adequate to meet the needs of the program yet still within the financial bounds allowed for by management, the

costs related to specific services and activities must be
identified. Generally, these costs are related to the
following items [26]:

Labor
* salaries – professional, clerical, and consultant
* social security payments
* unemployment and disability insurance taxes
* fringe benefits

Special Training Programs
* classroom
* hands-on refresher.

Materials
* capital expenditures – industrial hygiene, laboratory
  and safety equipment, supplies, and uniforms
* office supplies – desks, chairs, typewriters,
  computers, etc.
* replacement of expendable items
* depreciation of equipment and repairs

Overhead
* rent, lights, heat, gas, water, ventilation
* telephone, postage, freight
* fire and theft insurance
* repairs, alterations, maintenance, calibration
* laboratory services
* liability insurance
* medical, toxicological, and industrial hygiene
  computer information services

Employee Care
* health surveillance examinations – initial, annual,
  post-exposure, and exit
* emergency care services
* health and safety and other employee and advisory
  committees

Contingency Fund
* petty cash

Miscellaneous
* travel, including transportation, lodging and meals
* professional journal and textbooks
* special educational and professional advancement
  training

Other factors to consider in preparing a budget may include:

The need for new safety and health equipment to allow
for better and more efficient monitoring of the
employees and the workplace;
The addition of new health maintenance programs as
technology advances;

The addition of medical and associated services as
operations expand to new geographic areas; and
The addition of new personnel as the company expands.

Due to the differences in methods of cost accounting,
it is difficult, if not impossible, to quantitate the
average cost of a health and safety program for all
companies within the hazardous waste industry . One company
may charge the cost of certain overhead items to the health
and safety department, while other companies will exclude
these same items from operating costs. Therefore, how a
company handles its bookkeeping items will influence the
cost of the program. Generally, however, it is assumed that
of the budget needed to operate a health and safety program,
no more than 1.5 percent would be devoted to health and
safety staff functions, while no more than 1.0 percent of
the budget would be devoted to the various live organization
functions, including the needed training time [16].
One final word: it is important to recognize that
since the health and safety program is not a revenue
producing operation, it will be easy for management to
slight its budget when finances get tight; however, the
benefits of a preventative health and safety program will
lead to:

* increased employee productivity and efficiency;
* reduced absenteeism and illness;
* reduced workmen's compensation rates;
* reduced insurance premiums;
* reduced injury, severity, and frequency rates;
* reduced legal liabilities; and
* improved employee morale and involvement.

REFERENCES

1. Magnuson, E. "The Poisoning of America," Time Magazine, (September 1980), pp. 58-69.
2. National Environmental Protection Act, 1969.
3. Occupational Safety and Health Act, 1970.
4. Federal Water Pollution Control Act, 1970 and Amendments, 1972 P.L. 92-500.
5. Safe Drinking Water Act, 1974.
6. Clean Air Act, 1970 and Amended 1977.
7. Clean Water Act, 1977.
8. Resource Conservation and Recovery Act, 1976.
9. Toxic Substances Control Act, 1976.
10. Comprehensive Environmental Response, Compensation, and Liabilities Act, 1980.
11. Dalton, J. M., and T. F. Dalton. "Personnel Safety in Hazardous Material Cleanup Operations," in Proceedings of the 1980 National Conference on Control of Hazardous Materials Spills, (Nashville, TN: Vanderbilt University, 1980), pp. 264-269.
12. Finch, A. C. "Small Business Needs for Occupational Safety and Health Services," in Proceedings of Clinic Based Occupational Safety and Health Programs for Small Business, DHEW (NIOSH) Publication Number 77-172, Cincinnati, 1977, p. 23.
13. Kerr, L. E. "Impact of National Health Insurance on Occupational Safety and Health Services for Small Businesses," in Proceedings of Clinic Based Occupational Safety and Health Programs for Small Business, DHEW (NIOSH) Publication Number 77-172, Cincinnati, 1977, p. 6.
14. Bridge, D. P. "Developing and Implementing an Industrial Hygiene and Safety Program in Industry," American Industrial Hygiene Association Journal, 40(4):255-263 (1979).
15. McRae, A. D. and K. E. Lawrence, Eds. Occupational Safety and Health Compliance Manual, (Germantown, MD: Aspen Systems Corp., 1978).
16. Trauth, C. A., Jr., and J. B. Sorensen. "A New Approach for Assuring Acceptable Levels of Protection from Occupational Safety and Health Hazards," Sandia National Laboratories, SAND 81-1131C, May 1981, Albuquerque, NM.
17. Edward, S. "Quality Circles are Safety Circles," National Safety News, 127(6):31-325 (1983).
18. Griffin, R. E. "Safety Circles are 'The New Team in Town'," National Safety News, 127(6):31-35 (1983).
19. Halley, P. D. "Industrial Hygiene - Responsibility and Accountability," American Industrial Hygiene Association Journal, 41(9):609-615 (1980).

54

20. Gallagher, G. A. "Health and Safety Program for Hazardous Waste Site Investigations," paper presented to the New England Section of the Association of Engineering Geologists, Boston, MA, February 7, 1981.

21. Dahlstrom, D. L. "Working in Toxic/Hazardous Environments - A question of Health Surveillance," paper presented at the 184th National Meeting of the American Chemical Society, Kansas City, MO, September 12-17, 1982.

22. Dahlstrom, D. L. "Health and Safety Programs for the Hazardous Waste Worker," paper presented at the Engineering 1982 Conference, Buffalo, NY, February 1982.

23. "The Basics of Safety," Job Safety and Health - Bureau of National Affairs, Inc., Washington, D.C., April 26, 1983.

24. Gartseff, G. V., and D. L. Dahlstrom. "Safety Planning for Hazardous Waste Site Activities," paper presented at the 184th National Meeting of the American Chemical Society, Kansas City, MO, September 12-17, 1982.

25. Buecker, D. Ecology and Environment, Inc., personal communication, 1982.

26. Howe, H. "Organization and Operation of an Occupational Health Program," Journal of Occupational Medicine, 17(6):360-400 (1975).

4

INFORMATION GATHERING

Barrett E. Benson
Environmental Protection Agency
Office of Enforcement
National Enforcement Investigations Center
Denver, Colorado

The control of hazardous and toxic substances and the prevention of entry of these substances into the environment are regulated by the Toxic Substances Control Act (TSCA), the Resource Conservation and Control Act (RCRA), and the Comprehensive and Environmental Response, Compensation, and Liability Act (CERCLA or "Superfund"). These regulations require that wastes be identified and the degree of hazard assessed. Environmental response can be thought of as four categories of action, occurring in chronological order. These include:

1) characterization;
2) assessment;
3) mitigation; and
4) continuous monitoring.

Characterization is the first step in an environmental response to a hazardous waste problem. Data and information collected in this first step are interpreted in the second step to make decisions on how to accomplish the third. This chapter summarizes the methods for collecting available information and lists sources and data bases where the information can be found. The appendix at the end of the chapter lists selected data bases for hazardous waste investigations.

The investigation of a waste disposal site is aimed at answering the following questions:

* What is the waste composed of and what are its characteristics?
* Are the wastes adequately immobilized or destroyed at the site?

* By what routes can the wastes move off the site?
* What effects could occur (or might have occurred) through the discharge of wastes?
* If necessary, what steps can be taken to remedy the problem?

These questions should be pondered at all stages of investigation: the preliminary assessment, inspection and field investigation.

After a potential problem site has been identified, the first step is to collect all readily available information prior to conducting a site inspection. The following information should be gathered through telephone and personal contacts with knowledgeable persons, file searches, and analysis of aerial photographs.

1) Determine if an emergency exists;
2) Estimate potential severity of the problem and establish priorities for further investigation;
3) Focus the inspection and field investigation efforts on the proper areas;
4) Discover hazards to field personnel, allowing them to take proper safety precautions;
5) Incorporate whatever findings are available from previous studies of the site;
6) Develop an estimate of the kinds of resources needed to investigate the problem; and
7) Identify the parties responsible for the problems and obtain documentation to corroborate the identification.

SOURCES OF DATA

Contacts in Locale

Once a problem has been identified, the original source of that information, either private citizen or government official, should be asked to provide names of all other persons who might have knowledge about the site in question. If the contact is a private citizen, the names of anyone who might be able to corroborate the report should be requested. When appropriate, witnesses should be asked to prepare affidavits in support of their statements. If personal injury or property damage is claimed, ask for the name and telephone number of the attending physician or insurance adjuster. If the source of information is an employee of the facility under discussion, it is advisable to inform that person of the employee protection provisions of RCRA, Section 7001.

Government Files

After receiving a report of a possible waste disposal problem, the investigator should examine all available information in government files. Within the EPA regional offices, contacts in the Toxic Substances, Drinking Water, Solid Waste, and Enforcement programs should be consulted for information. Specifically, the investigator should determine if there is information on the site or facility which has been filed either through the RCRA or CERCLA notification process. State and local environmental agencies may have valuable information regarding sites, disposal practices, and other technical data. Information on whether the operator has an NPDES permit should be sought. If the operator has applied for this permit, the application forms may provide considerable information on the wastes disposed at the site and the engineering design of the facility. If no NPDES permits are held by the facility, then demonstration of discharges to surface waters may suffice for initiating a full enforcement investigation or enforcement action. In many cases, information may be available from state inventories of surface impoundments under the Safe Drinking Water Act (SDWA) or of open dumps under the Resource Conservation and Recovery Act (RCRA). In some cases, the U. S. Geological Survey (USGS) may have conducted investigations of groundwater pollution and found that the source is a waste disposal facility. Thus, USGS should be consulted for information on sites under study.

If the facility has applied for a state solid waste permit, a considerable amount of information may be available regarding geology, hydrology, and soils. Records of site visits and state enforcement actions should be requested. State water quality agencies may have data on ambient surface water and groundwater quality. Precise geographical location of the facility (geographical coordinates in degrees, minutes, and seconds) is important for obtaining aerial photography. In many cases, the county registrar of deeds will provide such information over the telephone; otherwise, they will generally respond to a written request. The investigator should also ask for highway direction to the site. Zoning or planning commissioners may be able to provide detailed maps of the site and its environs.

There are more than 150 computerized databases on a wide variety of subjects, including chemistry, toxicology, engineering, law business, and economics which can be accessed for information. Examples are:

1) Corporate information including subsidiaries, profit and loss statements, boards of directors, history of the company, products, etc..
2) Information on specific chemicals, such as toxicity, physical and chemical properties, manufacturers, and locations.
3) Ownership of property, operations, employees, etc.

As much information as possible should be compiled before using the automated system, including all the information available about the site and potentially responsible parties. In some cases, only the site name and location will be available. With the site name and location, personnel may be able to identify an owner, leasee, operator, etc.

1. DIALOG Information Retrieval Service

Over 200 databases are currently available in a variety of disciplines: chemistry, biology, energy, engineering, business, environmental law, economics, patents, and geology. The services also includes extracts of 10-K and 10-Q reports filed with the SEC. Another file, Chemical Regulations and Guidelines System, identifies all regulations which govern an individual chemical. The Congressional Information Service file is used to locate the testimony of witnesses at congressional hearings. For further information, call or write:

DIALOG Information Retrieval Service
3460 Hillview Avenue
Palo Alto, CA 94304
(800) 227-1960

2. SDC Search Service

Over 50 databases are available, many of which overlap with DIALOG files; this service also has a safety file and a paper chemistry file. For further information, call or write:

CDC Search Service
2500 Colorado Ave.
Santa Monica, CA 90406
(800) 421-7229

3. WESTLAW

Contains legal information, including full text of
cases back to 1932 for Supreme Court cases. U. S.
Court of Appeals and the U. S. District Courts are
searchable back to 1961. Also contains Shepard's
citations, which traces the history of a case to
locate all other cases that cite that case, and the
Forensic Services Directory which can be used to
locate possible expert witnesses. For further
information, call or write:

WESTLAW
50 W. Kellogg Blvd.
St. Paul, MN 55165
(800) 328-9833

4. JURIS (Justice Retrieval and Inquiry System)

Juris is a computerized legal information system
developed and operated by the Department of
Justice. It consists of Federal case law, Federal
statutes and regulatory material, Federal and state
case digest and attorney work products. (This
system is available only to Federal government
agencies.)

5. National Library of Medicine (NLM-MEDLARS)

This system contains databases such as Medline,
Toxline, Registry of Toxic Effects of Chemical
Substances (RTECS), Toxicology Data Bank (TDB), and
Cancerline - a main source for toxic and health
effects. For further information, call or write:

National Library of Medicine
MEDLARS Management Section
8600 Rockville Pike
Bethesda, MD 20209
(800) 638-8480

6. NIH-EPA Chemical Information System

Has four databases, including Oil and Hazardous
Materials Technical Assistance Data System (OHMTADS)
which contains data for any substance which has been

61

designated on oil or hazardous material; Subs-structure and Nomenclature Search System (SANSS), which is used for locating structural information in specific compounds; Mass Spectroscopy Search System (MSSS), used for obtaining Mass Spec information; and Federal Register Search System (FRSS), which contains citations to chemicals, substances, and materials published in the Federal Register since February 1, 1978. For further information, call or write:

CIS User Support Group
Computer Sciences Corp.
P.O. Box 2227
Falls Church, VA 22042
(800) 368-3432

7. Dun and Bradstreet

Dun and Bradstreet, a credit-reporting firm, provides business information reports for privately-owned companies. These reports include such information as business done by the company, company history, financial conditions, how it pays its bills, company locations, and principals of the company. For further information, call or write:

Dun and Bradstreet, Inc.
99 Church St.
New York, NY 10007
(212) 285-7000

8. Hazardline

The Hazardline system contains information on 2,350 substances which OSHA defines as hazardous in the workplace. This system is used to obtain OSHA standards, hazardous spill and waste disposal information and medical and first air information. For further information, call or write:

Occupational Health Services, Inc.
515 Madison Ave.
New York, NY 10022
(808) 223-8978

9. Defense Documentation Center

This system contains R&D reports produced by U. S.
Military organizations and their contractors. The
system is available only to Federal agencies.

In addition to the commercially available data systems,
the EPA has developed internal automated information
sources. Currently, these are available only to EPA. The
internal databases include the following:

1. Docket System

This system tracks all civil and criminal
enforcement actions taken by EPA or designated
states and all other enforcement actions initiated
against major violators of deadlines with regard to
the clean air and clean water pollution control acts.

2. Compliance Data System (CDS)

This file is used to track the completion of actions
for major sources of the five primary air pollutants.

3. Permit Compliance System (PCS)

This file is used to tract reporting requirements
under the FWPCA which prohibits any person from
discharging pollutants into a waterway from a point
source unless the discharge is authorized by a
permit. A list of reports that will be due during a
specified time is printed, and enforcement actions
can begin against permittees who are delinquent.

4. Consent Decree Inventory System

NEIC has the responsibility of developing this
system which will be capable of tracking consent
decree compliance status. About one-half of the
entered decrees are currently in the system.

5. Hazardous Waste Data Management System

This file contains data from regional files on
permits, enforcement, and inspections concerning
RCRA and Superfund actions.

6. Superfund Financial Assessment System (SFFAS)

The SFFAS is designed to calculate the ability of
the responsible party to pay for remedial action at

63

a hazardous waste site. The assessment should be used as a guide, along with the other factors, in determining final cleanup costs.

7. RCRA Financial Test

This system is currently in the planning stage. When implemented the system will analyze a company's financial data and be able to determine whether the facility passes the financial test to assure responsibility for closure and post-closure costs.

Records of Generators

RCRA requires that generators complete manifests for each shipment of hazardous wastes transported to a disposal, treatment or storage facility. Copies of these manifests and a biannual summary must be filed with the regional EPA office.

Treatment, storage, and disposal facilities must file an application for a permit in accordance with RCRA. These applications must detail: 1) the type and quantity of material handled or expected to be handled and; 2) additional information regarding the physical facilities. Permits will specify recordkeeping, monitoring and maintenance requirements, descriptions of the process, construction and design of disposal and storage facilities and contingency plans.

Approximately 35 states are now authorized to administer and enforce programs pursuant with RCRA in lieu of the Federal program. In such cases, states may institute operating requirements in addition to those designated by RCRA.

The office of solid waste maintains a data base of RCRA information and can make this information available to the EPA regional offices. Much of this data is accessible by computer.

DATA NEEDS

Geology

Information on local bedrock types and depths may be important to an investigation of groundwater pollution

64

problems, particularly where producing aquifers lie beneath
the water table aquifer. Sedimentary strata (limestones,
sandstones, and shales) tried to channel groundwater flows
along bedding planes and flow directions may be determined by
the dip in the strata. Where limestones are present in humid
climates, solution channels may develop allowing very rapid
transport of pollutants over long distances with little
attenuation. Igneous and metamorphic bedrock (granites,
diorite, marble, quartzite, slate, gneiss, schist, etc.) may
permit rapid transport of polluted groundwater along fracture
zones. Depth of bedrock may be an important factor in
selecting the appropriate type of remedial action. Sources
of this information include USGS reports and files, state
geological survey records, and local well driller logs.

Soils and Overburden

Soil and overburden types and permeabilities are very
important factors in evaluating the pollution potential of a
waste management site. Highly permeable soils (i.e.,
10^{-3} cm/sec) may permit rapid migration of pollutants, both
vertically and horizontally, away from containment areas.
Rates of attenuation of pollutants in the unsaturated zone
and underlying aquifers are a function of soil chemistry and
physical characteristics. It is often important to ascertain
the availability and quality of local clays in considering
possible remedial measures. The USGS, the Soil Conservation
Service, Agricultural Extension Service Agents, state
geological survey records, local well drillers, and local
construction engineering companies may be able to supply such
information.

Climate

Locale climate may also be an important factor in the
pollution potential of a facility. Mean values for
precipitation, evaporation and evapor-transpiration, and
estimated infiltration can be used as general indicators of
the potential for groundwater pollution at a site. In many
cases groundwater or surface water pollution has occurred due
to unusually high amounts of precipitation. Even in an arid
region where little or no recharge to groundwaters generally
occurs, an extremely wet year may create a serious pollution
problem. In evaluating the pollution potential of a

"non-discharging" surface impoundment, it may be expedient to calculate a mass balance to determine if seepage through the walls or bottom is occurring. For these purposes, the investigator would find it necessary to consult monthly or seasonal precipitation and evaporation (or temperature) records during the period of operation. The maximum recorded or estimated rainfall for a given period of time (24 hours, 48 hour or monthly) may be an important factor in evaluating the amount of freeboard needed in a surface impoundment. Where airborne contaminants may be a problem, it will be important to determine prevailing wind patterns and velocities. Such information can be obtained through:

National Climatic Center
Department of Commerce
Federal Building
Asheville, North Carolina 28801
(FTS) 672-0683
(204) 258-2850

Geohydrology

In most cases, the investigator will need to acquire information on the groundwater hydrology of a site and its environs. Depths to the water table and any underlying aquifers, characteristics of confining layers, piezometric surfaces (heads) of confined aquifers, direction(s) of flow, existence of perched aquifers, and areas of interchange with surface waters will be vital in evaluating the pollution potential of a facility. Groundwater use in the area of the site should be thoroughly investigated to find the depths of local wells, pumping rates, and the ways in which the water is utilized. Sources of such information include the USGS, state geological surveys, local well drillers, and state and local water resources boards. A list of all state and local cooperating offices is available from the USGS Water Resources Division in Reston, Virginia 22092. This list has also been distributed to EPA Regional Offices. Water quality data, including surface waters, is available through the USGS via their automated NAWDEX system; for further information: telephone FTS: 928-6081 or (703) 860-6031.

Surface Wastes

The locations of all surface waters or dry stream beds in the area should be investigated, and surface gradients on and around the site should be determined. If surface waters down-gradient from the site are used for drinking, recreation, fishing, irrigation, or livestock watering, this should be noted. Where pollution of surface waters is suspected, it is advisable to collect available baseline water quality data and stream flow rates. In addition, it may be necessary to get information on all NPDES permitted discharges by other operations in the vicinity of the site under investigation. Topographic maps, aerial photography, and the NAWDEX systems (see above) of USGS can provide useful information on surface waters.

Sensitive Environments

The investigator also needs to determine if the site is located in a sensitive environment; e.g., in a floodplain, inside or adjacent to wetlands, in a recharge zone of a sole source aquifer, in an area of karst topography, or in a fault zone. In general, the potential for long-term environmental disruption must be determined, if a discharge of hazardous wastes should occur.

Site Environs

Before conducting a site visit, it is advisable to gather some information about the area surrounding the site. For the safety of those conducting a site visit, the names and telephone numbers of the police and fire department responsible for that area should be noted. Furthermore, these departments may provide information on violations of laws and safety codes, including records of past incidents at the site under investigation. Sources of drinking water supplies in the area, both public and private, should be noted. In addition, the investigator should check to see what analyses have been performed on these water supplies and request copies of all data. It may be important to find out what kind of treatment system is used by the public water supply, it is important to determine the locations of water mains -- these could be conduits for entry of polluted

groundwater into the public water system. Likewise,
information should be gathered on the local sewer and storm
drain systems to determine possible infiltration or illegal
discharge points. The most important characteristics for
determining the hazards in a given situation are population
densities and distances to residences, schools, commercial
buildings and any other facilities in the vicinity of the
waste site which may be occupied. The investigator should
also try to determine if any flammables or explosives, such
as liquefied natural gas, are stored near the site. General
land use of the environs should be studied; if crops or
livestock are raised in the area, the types should be noted.
It may be important to seek information on the wildlife or
aquatic life in the area.

Hazardous Waste Management Information

     If no information on wastes is available from government
sources, it may be necessary to proceed with the site
inspection and field investigations without benefit of
background information. In some cases, however, it may be
possible to form a hypothesis on the kinds of waste present
at the facility. Where a landfill contains both municipal
and industrial wastes, it is probable that much of the waste
comes from the local industries. If approximate dates of
operation of the facility are known, local officials or the
Chamber of Commerce may be able to provide information on
industries operating locally during that time period. In the
case of an on-site (at the generator's site) facility, it may
be possible to determine the type of waste present.
Information on the composition of waste streams associated
with various industrial processes may be obtained from the
EPA Hazardous and Industrial Waste Division of the Office of
Solid Waste in Washington, D.C. Other sources are the EPA
Assessment of Industrial Hazardous Waste Practices
(14 industries covered) and Pollution Control in the Organic
Chemical Industry (Noyes Data Corporation). In gathering
such information, the "Hazardous Waste Site Survey Record"
and the "Hazardous Waste Site Identification and Preliminary
Assessment" forms should serve as models for the kinds of
questions to be asked.

Available Aerial Photographs

Aerial reconnaissance is an effective and economical tool for gathering information on waste management sites. For this application, aerial reconnaissance should include aerial photography gathering data during daylight hours. The thermal infrared scanning is occasionally done during daylight, but has greater utility at night.

In general, reconnaissance might be performed during preliminary investigation to obtain data on the following:

* Approximate amount of solid and liquid waste disposed on various sites;
* Casual disposal sites (sites in which control measures are lax or do not exist);
* Illegal or promiscuous dumping within industrial, state, or municipal facilities or at remote sites;
* Unauthorized disposal of liquid waste at solid waste disposal sites;
* Visible environmental effects resulting from disposal practices such as spills, surface runoff patterns, surface leachate flow, impoundment leakage, and damaged or stressed vegetation in the immediate environs of disposal sites;
* Visible environmental effects resulting from disposal practices such as spills, surface runoff patterns, surface leachate flow, impoundment leakage, and damaged or stressed vegetation in the immediate environs of disposal sites;
* Superficial geology such as faults in or near the site;
* Storage container location;
* Container inventories;
* Waste disposal sites not directly visible or readily accessible from the ground;
* Facility design and operation, pertinent to the investigation;
* Land use of site environs; and
* Location of possible hazards to inspectors.

Information on aerial reconnaissance and the aerial data processing is available at these EPA offices:

National Enforcement Investigation Center (NEIC)
Building 53, Box 25227
Denver, Colorado 80225
(FTS) 234-4650

Environmental Photographic Interpretation Center (EPIC)
P.O. Box 1587
Vint Hill Farm Station
Warrenton, Virginia 22186
(FTS) 557-3100

or

Environmental Monitoring Systems Laboratory (EMSL-LV)
P.O. Box 15027
Las Vegas, Nevada 89114
(FTS) 545-2660

Each office maintains its own archive of aerial
reconnaissance imagery (photography and thermal scanner
data). EPIC has access to historical imagery from other
government agencies.

Archival Imagery

Federal agencies have been using aerial photography for
a variety of purposes for several decades. Usable
photographs less than five years old will usually be
available for a site. Frequently, however, the scale will be
too small to observe details of the site without considerable
magnification of the imagery. In cases where it is important
to gather information on the locations, areal extent, and
historical development of facility operations (e.g., the size
and locations of old landfill cells), archival photography
can prove invaluable.

Archival photographs are available from the following
sources:

National Cartographic Information Center
U. S. Geological Survey
Building 25, Denver Federal Center
Denver, Colorado 80225

and

EROS Data Center
Sioux Falls, South Dakota 57102

Photographs taken before 1950 are available from the
National Archives. Generally, the requester must specify the
geographical coordinates (latitude and longitude) of the site

when requesting photography. Information on the photography
available for a given site can usually be obtained through
the above facilities in less than 30 minutes. Standard
orders for copies of photographs are processed within 6
weeks; priority requests require approximately 1 week at a
significantly higher cost. Photo interpretation is available
through NEIC, EPIC and EMSL (LV).

Preparation of Sketch Map

The sketch map should utilize available aerial
photographs of the site and reflect any updated information
collected during the preliminary assessment. In some
instances, the aerial photographs will be readily available
and of sufficient quality to be used instead of a sketch
map. It is often convenient, however, to have multiple
copies of an easily reproducible site sketch map which can be
marked and drawn on as the investigation progresses.

EVALUATION OF PRELIMINARY DATA

After all of the data and information have been
gathered, to complete the preliminary assessment, an initial
evaluation of the information should be performed after all
of the preliminary data and information have been gathered.
The information should be used to evaluate the following:

* the existence (or nonexistence) of a potentially
  hazardous waste problem;
* probable seriousness of the problem and the priority
  for further investigation or action; and
* the type of action or investigation appropriate to the
  situation.

If the available information is not sufficient to determine
the above, a field reconnaissance may be required.
It is important to remember that information gathering
is an ongoing process during all phases of a hazardous waste
site investigation. Information discovered later or after
the initial gathering process may provide leads to other
existing data that were previously unknown. It is essential
to keep detailed records and notes of what is learned during
each step of identification. All further activities,
including reconnaissance and actual field investigations,
should aid in collecting information.

A dominant concern during all phases of a hazardous waste site characterization is safety, for the general public, site investigators, and remedial action workers. A comprehensive health and safety program for on-site investigations and subsequent remedial work at hazardous waste sites is required to protect investigators, workers, the public and environment. The program should provide general guidance for on-site investigations to assure the highest level of protection for the unique characteristics of each facility on-site. Each individual involved with the site is responsible for complying with the health and safety program established for the hazardous waste site investigations.

All individuals scheduled to visit hazardous waste sites are required to participate in a medical surveillance program. A comprehensive medical and occupational history, examination and laboratory workup must be completed on all individuals prior to on-site activities.

Proper training is the most important factor in implementing a successful health and safety program. Worker awareness of health and safety hazards and a personal commitment to safety procedures and emergency responses are what ultimately help them and the general public avoid adverse health effects due to exposure to contaminants. All individuals are required to participate in a health and safety training program prior to working on-site. At a minimum, this training should cover the following:

* special chemical and physical hazards and potential health effects at a site;
* first aid;
* emergency and routine communications;
* rescue operations;
* evacuation procedures;
* personal protective clothing and equipment use and maintenance;
* decontamination;
* fire fighting; and
* hands-on training in simulated sites.

Prior to entering a site for the first time, the site must be checked for radioactivity, explosivity, and oxygen content. If the oxygen content is less than 19.5 percent, self-contained breathing apparatus must be used. Normal background radioactivity is approximately 0.2 milli-Roentgens per hour (mR/hr); however, activity of 10.0 mR/hr may be acceptable for the period of exposure associated with a site investigation. If explosivity readings greater than 20 percent of the Lower Explosive Limit (LEL) are detected, a very careful survey of the area must be made. Readings approaching or exceeding 50 percent of the LEL are cause for immediately withdrawing personnel and notifying the fire department.

In cases of immediate hazard to investigators, workers or the public, any individual on the scene should take all practicable steps to eliminate or neutralize the hazard; this may include leaving the site. The supervisor or on-scene coordinator must take the necessary action to ensure that the work can be completed safely. Such steps may include changes in procedure, removal or neutralization of a hazard, consultation with appropriate experts, or bringing in specialists such as Explosive Ordinance Disposal units.

The information and data gathered prior to a site investigation can be used to determine if there is an immediate hazard present or if one will be created by remedial actions. The information must be included in the study plan for the investigation or made available to all individuals participating in the investigation or remedial actions. Worker awareness, compliance with safety programs, and adherence to approved technical procedures for investigations and remedial actions are the most important aspects of protecting people and the environment. The USEPA, Environmental Monitoring Systems Laboratory, Las Vegas, published Characterization of Hazardous Waste Sites -- A Methods Manual, Vol. I, Integrated Approach to Hazardous Waste Site Characterization, 1984, which details the methods used by the EPA and its contractors in conducting Remedial Investigations/Feasibility Studies (RI/FS).

Data Base Name	Subject Coverage
AEROS	Aeromatic and Emissions Reporting System (AEROS)is a management information system for EPA's air pollution research and control programs. Eleven retrievel systems are sub-systems of AEROS system:

   1. National Emissions Data System (NEDS).

   2. Storage and Retrieval of Aerometric Data (SAROAD) System.

   3. Hazardous and Trace Substance Emissions System (HATREMS).

   4. Source Test Data (SODAT) System.

   5. State Implementation Plans (SIPS) regulation system.

   6. Emissions History Information System.

   7. Air Pollutant Emissions Report (APER) tracking system.

   8. Weighted Sensitivity Analysis Program (WSAP).

   9. Source Inventory and Emission Factor Analysis (SIEFA) Program.

10. Computer Assisted Area Source emissions (CAASE) Gridding System – Performs calculations to apportion NEDS county emissions totals to subcounty gridded areas.

11. Regional Emissions Projection System (REPS).

Appendix I. Selected Data Bases for Hazardous Waste Site
Investigations

Coverage Dates	Update Frequency	Sponsoring Agency	Comments
Current	Continuous	EPA Research Triangle Park	Data and Computer Programs

Data Base Name	Subject Coverage
AGRICOLA	Covers worldwide journal and monographic literature in agriculture and related subject fields, including general agriculture and rural sociology; animal science; forestry and plant-related areas; entomology; and agricultural engineering. Includes agriculture Canada.
APTIC	Covers most sources for citations concerning all aspects of air pollution, its effects, prevention and control.
ASI	American Statistics Index covers statistical publications containing the entire spectrum of social economic and demographic data collected and analyzed by all branches and agencies of the U.S. government.
BIOSIS PREVIEWS	Covers all aspects of the life sciences, drawing upon all original published literature for citations. Corresponds to Biological Abstracts and Biological Abstracts/RRM.
CA SEARCH	Covers all aspects of the Chemical Abstracts Service.
CA CONDENSATES/ CASIA	Gives general subject index headings and CAS registry numbers for documents covered by CA condensates.
CAB ABSTRACTS	Comprehensive file of agricultural information pertaining to all significant material and covering every aspect of crop and animal science.

Appendix I. Selected Data Bases for Hazardous Waste Site
Investigations--continued

Coverage Dates	Update Frequency	Sponsoring Agency	Comments
1970-pres.	Monthly	U.S. National Agricultural Library	Over 1.9 million records; Citations only
1966-Oct. 1978	----	Manpower and Technical Info. Branch, EPA	89,000 citations; Abstracts
1973-Pres. (some material from 1960's)	Monthly	Congressional Information Service, Inc.	Over 94,000 records; Abstracts
1969-Pres.	Biweekly	Information Service	4,214,000 records;
1967-Pres.	Biweekly	Chemical Abstracts Service	6,395,000 records;
1972-Pres.	Biweekly	Chemical Abstracts Services	Corresponds to CA files; Description and identifiers only
1973-Pres.	Monthly	Commonwealth Agricultural Bureau	Over 1,618,000 items; Abstracts

Appendix I. Selected Data Bases for Hazardous Waste Site
Investigations--continued

Data Base Name	Subject Coverage
CANCERLIT	(Formerly Cancerline.) Contains information on various aspects of cancer taken from over 3,000 U.S. and foreign journals as well as selected monographs, papers, reports and dissertations.
CHEMDEX	Chemdex is based on the CA Registry Nomenclature File, which is a repository for names associated with substances that have been registered by Chemical Abstracts. In addition to CA's rigorous nomenclature data, this file contains registry numbers, molecular formulas, synonyms and ring system information.
CHEMLINE	Chemline is an online chemical dictionary file providing a mechanism for searching and retrieving chemical substance names. It contains records for chemicals that are identified by chemical abstracts service registry numbers and are cited in either Toxline, Toxback, TDB, or RTECS.
CHEMNAME	Contains CAS registry numbers, CA substance index names, molecular formulas, chemical name synonyms and periodic classification terms for chemical substances.
CLAIMS/CHEM	U.S. chemical and chemically-related patents plus some foreign equivalents
COMPREHENSIVE DISSERTATION ABSTRACTS	Interdisciplinary listing of almost all doctoral dissertations accepted since 1861 by accredited degree granting institutions in the U.S. plus some non-U.S. universities.
CONFERENCE PAPERS INDEX	Covers approximately 1,000 scientific and technical meetings worldwide and the 100,000 papers presented.

Appendix I. Selected Data Bases for Hazardous Waste Site
Investigations--continued

Coverage Dates	Update Frequency	Sponsoring Agency	Comments
1963–Pres.	Monthly	National Cancer	Over 100,000 Abstracts of Articles
Since 1972	Quarterly	Chemical Abstracts Serv. of the Amer. Chemical Society	694,461 substances
----	Irregular Updates	National Library of Medicine	1,000,000 names; 525,000 substances
1967–Pres.	Quarterly	Chemical Abstracts Serv. & Lockheed	Over 1,766,000 substances
1950–70		IFI/Plenum Data Co.	265,000 citations
1861–Pres.	Monthly	Xerox Univ. Microfilms	Over 845,000 citations; Citations only
1973–Pres.	Monthly	Data Courier	1,061,000 records

Appendix I. Selected Data Bases for Hazardous Waste Site
Investigations--continued

Data Base Name	Subject Coverage
EDB	The energy data base covers all information of interest to DOE in almost every area of research.
EIS INDUSTRIAL	Information on 130,000 establishments operated by 67,000 firms with current annual sales of over $500,000, describing employment, sales, market share and production.
EMI (EMIC)	Environmental mutagens--information concerning chemical mutagen research.
ENVIRONMENTAL PERIODICALS BIBLIOGRAPHY (EPB)	Covers the very broad field of general human ecology, atmospheric studies, energy, land resources, water resources and nutrition and health from 300 periodicals.
EXCERPTA MEDICA	Covers all fields of medicine plus extensive coverage of the drug and pharmaceutical literature and other areas such as environmental health and pollution control.
FEDERAL INDEX	Substantive comments from the Congressional Record, Federal Register, Presidential documents, and the Washington Post. Trends and developments in Washington are provided by citations to the Code of Federal Regulations, the US Code, public laws, congressional bills, and resolutions and reports. Coverage extends to proposed rules, regulations, bill introductions, speeches, hearings, roll calls, reports, vetoes, court decisions, executive orders, contract awards, etc.
FEDERAL REGISTER	Contents correspond to the Federal Register Abstracts

Coverage Dates	Update Frequency	Sponsoring Agency	Comments
1974-Pres. (contains material back to late 1800s)	5,000 items semimonthly	DOE Technical Information Center	98,700 citations; abstracts (after 6/1/76)
Current	Replace 3 times/year	Economic Information Systems	161,000 records; Citations
1976-Pres.		DOE-TIC	Abstracts
1973-Pres.	Bimonthly	Environmental Studies Institute	Over 254,000 records; Citations
June 1974-Pres.	Monthly	Excerpta Medica	3,067,000 records; Citations
Oct. 1976-Pres.	Monthly	Predicasts	213,000 citations; Citations only
March 1977-Pres.	Weekly	Capitol Services	211,000 citations

_____

Data Base Name	Subject Coverage
GEOARCHIVE	Covers geoscience information. Mineral and petroleum production and resources and new names typify the dta currently added to the fields of geophysics, geochemistry, geology, paleontology and mathematical geology.
GEOREF	Covers geosciences literature from 3,000 journals, plus the geosciences conferences and major symposia and monographs in all areas of the geosciences.
IPA	International Pharmaceutical Abstracts information on all phases of development and use of pharmaceuticals.
MEDLINE	Bibliographic citations to worldwide medical literature corresponds to Index Medicus.
MGA	Covers meterorological and geoastrophysical research published in both foreign and domestic literature. Based on Meterorological and Geoastrophysical Abstracts.
NAWDEX	National Water Data Exchange--Contains information concerning water data availability, source and some major data characteristics.
NBI	National Biomonitoring Inventory--Information on on-going biomonitoring projects in the U.S.

Appendix I. Selected Data Bases for Hazardous Waste Site
             Investigations—continued

Coverage Dates	Update Frequency	Sponsoring Agency	Comments
1969-Pres.	Monthly	Geosystems of London	519,000 citations; Abstracts
1961-Pres.	4,000 records/ month	American Geological Inst.	939,000 items; Citations
1970-Pres.	500-600 added monthly	Amer. Society of Hospital Pharmacists	93,000 reports
1976-Pres.	Monthly	National Library of Medicine	4,415,000 citations; Abstracts
1970-Pres.	Irregular updates	American Meteorological Society	103,500 records;
1700's to Pres.	As necessary	U.S. Geological Survey	Data from 61,500 sites stored
Current	As necessary	DOE-TIC	Abstracts

Appendix I. Selected Data Bases for Hazardous Waste Site
Investigations--continued

Data Base Name	Subject Coverage
NCC	National Climatic Center--contains historical and current weather information and related data. The data is generated by: NOAA's Weather Serv.; the U.S. Navy and U.S. Air Force weather services; the Federal Aviation Administration; the U.S. Coast Guard; and cooperative observers.
NRC	The National Referral Center file is a non-bibliographic file containing descriptions of organizations qualified and willing to answer questions on virtually any area of science and technology, including the social sciences.
NSA	The Nuclear Science Abstracts--subjects scope includes all nuclear science and technology.
NSC	Covers all pertinent literature on nuclear safety information.
NSR	The Nuclear Structure Reference data base contains: 1. The entire contents of "Nuclear Structure References, 1969-1974," supplement to Vol. 16, Nuclear Data Sheets. 2. Complete contents of the 1975 "Recent References" issues of Nuclear Data Sheets; and 3. References to reports and informal communications (secondary sources) received by the nuclear data project during the years 1973-1975.
NTIS	Complete government reports announcement file. Contains abstracts of research reports from over 240 government agencies including NASA, EPA, HEW, etc.

Appendix I. Selected Data Bases for Hazardous Waste Site
Investigations—continued

Coverage Dates	Update Frequency	Sponsoring Agency	Comments
1800's-Pres.	Continuous	National Oceanic and Atmospheric Admin.	Data sheets
Current	Continuous	National Referral Center for Science & Technology	Citations only
1967-1976	Closed	DOE-TIC	554,597 records; Abstracts
1963-Pres.	1,000 citations per month	Nuclear Safety Infor. Ctr., Oak Ridge Nat'l Laboratory	101,340 items; Abstracts
1974-Pres.	5,000 entries per year	Oak Ridge Nat'l Laboratory	30,236 items; Citations only
1964-Pres.	biweekly updates	National Technical Infor. Serv.	1,053,000 citations; Abstracts

Data Base Name	Subject Coverage
OHM-TADS	Oil and Hazardous Materials-Technical Assistance Data System contains data on materials that have been designated oil or hazardous materials. The system is designed to provide technical support for dealing with potential or actual dangers resulting from the discharge of oil or hazardous substances.
PARCS also (PPIS)	A centralized source of information on all pesticides registered by EPA. An updated file National Pesticide Information Retrieval System (NPIS) is being developed.
POLLUTION ABSTRACTS	Corresponds in coverage to the printed abstracts publication. Covers foreign and domestic reports, journals, contracts and patents, symposia, and government documents in the areas of pollution control and research; specifically, water, marine, land and thermal pollution; pesticides; sewage and waste treatments; and legal developments.
RASS	Rock analysis Storage System--contains information on samples submitted for analytical work. Information includes location, formation, sample name, age, description, economic geology, data, and geochemical data.
RING DOC	RINGDOC covers over 400 of the world's scientific journals to provide extensive coverage of the pharmaceutical literature. Access points to the citations include keywords and multipunch coded data (representing chemical fragments).
RTECS	Registry of Toxic Effects of Chemical Substances.

Appendix I.   Selected Data Bases for Hazardous Waste Site
Investigations--continued

Coverage Dates	Update Frequency	Sponsoring Agency	Comments
Oct. 1978–	Continuous	EPA–Oil & Special Materials Controls Division	Data
----	Continuous	EPA	Data
1970–Pres.	Bimonthly	Data Courier, Inc.	102,500 citations; Citations only
1962–Pres. 1968–Pres. (2 files)	As necessary	U.S. Geological Survey	135,000 records; 292,000 records
1964–Pres.	10,000 items a month	Dewent Publications, Ltd.	466,000 items
1900–Pres.	Quarterly	NIOSH	69,000 records

Data Base Name	Subject Coverage

SAFETY

Safety provides international coverage of the literature in 6 major areas: general safety, industrial and occupational safety, environmental and ecological safety, and medical safety.

SCISEARCH

Multidisciplinary index to the literature of science and technology.   Based on Science Citation Index which indexes approximately 2,600 major scientific and technical journals.

STORET

Storage and retrieval of water quality data--repository for water quality data that contains records of water quality parametric data by sampling site.

TOXICOLOGY DATA BANK (TDB)

Contains facts and data from some 80 standard references textbooks, handbooks, monographs, and criteria documents, for approximately 4000 substances; 1100 of these have been completed; 1500 are in process.   TDB contains approximately 60 different categories of data, such as chemical, physical, biological, pharmacological, toxicological, and environmental facts.

TOXLINE

Contains data on toxicity and adverse effects of environmental pollutants and chemicals on the human food chain, laboratory animals, and biological systems; and analysis techniques in the following subfiles:
CBAC--1965-76:   Chemical abstracts, biochemistry sections.
EMIC--1971-74:   Environmental Mutagen Information Center.
ETIC--1950:   Environmental Teratology
HAYES--1970:   EPA Pesticide File

Appendix I. Selected Data Bases for Hazardous Waste Site
Investigations--continued

Coverage Dates	Update Frequency	Sponsoring Agency	Comments
June 1975-Pres.	Bimonthly	Cambridge Scientific Abstracts, Riverdale, MD	Abstracts
1974-Pres.	Monthly	Institute for Scientific Information	5,568,000 records; Citations only
1976-Pres.	Bimonthly	Paint Research	Abstracts
1978-Pres.	Continuous	National Library of Medicine	4000 Substances
Varies with subject file	Varies with subject file	National Library of Medicine	479,926 records; Citations; (Abstracts available)

Data Base Name	Subject Coverage
TOXLINE--continued	HEEP--1971-76: Health Effects of Environmental Pollutants IPA--1970-76: International Pharmaceutical Abstracts PESTAB--(Formerly HAPAB)--1966-76: Pesticide Abstracts, EPA TERA--1971-74: Teratology TMIC--1971: Toxic Materials Information Center TOXBIB--1968-76: Index Medicus Toxicity Subset
WATSTORE	Water Data Storage and Retrieval System--contains data on the occurrence, quantity, quality, distribution, and movement of surface and underground waters.
WESTLAW	Contains Supreme Court full text and headnote summaries from 1932-present; headnote summaries for all reported Federal court cases from 1960-present; and all reported State Appellate Court cases from 1967-present. Full test accessible for all Federal court cases in the data base and Appellate Court cases from 1977-present. The subfiles correspond to the units of the West Company National Reporter System.
WRA	Corresponds to the semimonthly abstracting journal, Selected Water Resources Abstracts. Covers the water-related aspects of the life, physical, and social sciences as well as related engineering and legal aspects of the characteristics, conservation, control, use or management of water. Input material for the abstracts comes from selected organizations with active water resources research programs which are supported as "Centers of Competence."

Appendix I. Selected Data Bases for Hazardous Waste Site
Investigations--continued

Coverage Dates	Update Frequency	Sponsoring Agency	Comments
Historical to Pres.	As necessary	U.S. Geological Survey	Includes data on over 100,000 sites
1932-Pres.	Continuous	West Publishing Company	Number of items varies with subject files
1969-Pres.	1,000 items a month	Water Resources Scientific Information Center	170,000 items; Abstracts

REFERENCES

1. Dictionary of Geological Terms, American Geological Institute 1962. (Garden City, NY: Doubleday and Co.) 545 p.
2. Baedecker, M. J., and W. Back. "Hydrogeological Processes and Chemical Reactions at a Landfill," Groundwater 17(5):429-437 (1979).
3. Bentall, R. "Shortcuts and Special Problems in Aquifer Tests," U.S. Geological Society Report, Water Supply Paper 154-C, Government Printing Office (1963).
4. "Ground Water Manual," U.S. Bureau of Reclamation, Government Printing Office (1977).
5. "Water Measurement Manual," 2nd ed., U.S. Bureau of Reclamation, Government Printing Office (1967).
6. Butson, K. D., and W. L. Hatch. "Selective Guide to Climatic Data Sources," National Climatic Center, Asheville, NC, Key to Meteorological Records Documentation No. 4.11, 142 p.
7. Campbell, M. D., and J. H. Lehr. Water Well Technology. (New York: McGraw-Hill Book Co., 1973).
8. "Federal Environmental Data: A Directory of Selected Sources, National Technical Information Service, Springfield, VA, Capital Systems Group, Inc. (Nov. 1977).
9. Casarett, L. J., and J. Doull, Eds. Toxicology (New York: Macmillan Publishing Co., 1975.
10. Clarke, J. H., Ziegler, F. G., Tennant, D. S., Harbison, R. D. and R. C. James. A Model for Assessment of Environmental Impact of Hazardous Materials Spills and Leaching. (Nashville: Recra Environmental and Health Sciences, Inc., 1980).
11. Clarke, P. F., Hodgson, H. E. and G. W. North. A Guide to Obtaining Information from the U.S.G.S., 2nd ed., U.S. Geological Survey, Arlington, VA, Circular 777 (1979).
12. Compton, R. R. Manual of Field Geology. (New York: Wiley and Sons, 1967).
13. Coperhaver, E. D. and B. K. Wilkinson. "Movement of Hazardous Substances in Soil: A Bibliography, Volume 1: Selected Metals," EPA-600/9-79-024 a (Aug. 1979).
14. Copenhaver, E. D. and B. K. Wilkinson. "Movement of Hazardous Substances in Soil: A Bibliography, Volume 2: Pesticides," EPA-600/9-79-0246 (Aug 1979).
15. "Soil Sampling," Department of the Army, Army Corps of Engineers, EM 1110-2-1907 (1972).
16. Davis, S. N. and R. J. M. DeWiest. Hydrogeology, (NY: Wiley and Sons, May 1967, 463 p).
17. Deichman, W. B. and H. W. Gerarde. Toxicology of Drugs and Chemicals, (NY: Academic Press, 1969, 805 p).

18. Departments of the Army and Air Force. <u>Well Drilling Operations</u>, (Worthington, OH: National Water Well Assoc., 188 p).

19. De Vera, E. R., Simmons, B. P., Stephens, R. D., and D. L. Storm. <u>Samplers and Sampling Procedures for Hazardous Waste Streams</u>, (Cincinnati: Environmental Protection Agency, EPA-600/2-80-018, 70 p).

20. Environmental Protection Agency. "Proceedings of the Sixth Annual Research Symposium on Disposal of Hazardous Waste at Chicago, Ill.," (Cincinnati: EPA-600/9-80-010, March 17-20, 1980, 291 p).

21. Everett, L. G. and E. W. Hoylman. "Groundwater Quality Monitoring of Western Coal Strip Mining: Preliminary Designs for Active Mine Sources of Pollution," (Las Vegas: EPA-600/7-80-110, 105 p).

22. Evett, J. B. <u>Surveying</u>, (NY: Wiley and Sons, 1979, 273 p).

23. Fenn, D., Cocozza, E., Isbister, J., Braids, O., Yore, B., and P. Roux. "Procedures Manual for Groundwater Monitoring at Solid Waste Disposal Facilities," (Cincinnati: EPA 530/SW-611, 169 p).

24. Freeze, R. A. and J. A. Cherry. <u>Groundwater</u>, (Englewood Cliffs, NJ: Prentice-Hall, 1979, 604 p).

25. Frey, D. G., ed. <u>Limnology of North America</u>, (Madison, WI: University of Wisconsin Press, 1963, 734 p).

26. Fuller, W. H. "Investigation of Landfill Leachate Pollutant Attenuation by Soils," (Cincinnati: EPA-600/2-78-158, 1978, 219 p).

27. Gale Research Company. <u>Climates of the States</u>, (Detroit: Book Tower, 1978).

28. Gary, M., McAfee, R. and C. L. Wolf, eds. <u>Glossary of Geology</u>, (Falls Church, VA: American Geological Inst., 1977, 805 p).

29. Geraghty and Miller, Inc. "Surface Impoundments and their Effects on Groundwater Quality in the United States - A Preliminary Survey," (Washington: EPA 570/9-78-004, June 1978, 276 p).

30. Geswein, A. J. "Liners for Land Disposal Sites, An Assessment," (Washington: EPA 530/SW-137, March 1975, 66 p).

31. Gibb, J. P. and R. A. Griffin. "Groundwater Sampling and Sample Preservation Techniques (1st Annual Report)," (Cincinnati: EPA, 1979).

32. Gibb, J. P., Schuller, R. M. and R. A. Griffin. "Monitoring Well Sampling and Preservation Techniques," <u>Proceedings of the Sixth Annual Symposium on Disposal of Hazardous Wastes,</u> Chicago: March 17-20, 1980. (Cincinnati: EPA-600/9-80-010, March 1980, pp. 31-38).

33. Gilluly, J., Waters, A. C. and A. O. Woodford. <u>Principles of Geology</u>, 3rd ed., (San Francisco: Freeman and Co., 1968, 687 p).

34. Greenwood, D. R., Kingsbury, G. L. and J. G. Cleland. "A Handbook of Key Federal Regulations and Criteria for Multimedia Environmental Control," (Washington: EPA 600/7-79-175, Aug. 1979, 272 p).

35. Gosselin, R. E., Hodge, H. C., Smith, R. P. and M. N. Gleason. Clinical Toxicology of Commercial Products, 4th ed., (Baltimore: Williams and Wilkins, 1977, 799 p).

36. Griffin, R. A. and N. F. Shimp. "Attenuation of Pollutants in Municipal Landfill Leachate by Clay Minerals," (Cincinnati: EPA 600/2-78-157, Aug. 1978, 147 p).

37. Hammer, M. J. Water and Waste-Water Technology, (NY: Wiley and Son, 1975, 502 p).

38. Harding, S. T. "Lakes," in Hydrology, O. C. Meinzer, ed. (NY: Dover Publications, 1942, pp. 220-242).

39. Johnson, A. I. "An Outline of Equipment Useful for Hydrologic Studies," (Denver: U.S. Geological Survey, Open-File Report, 1964, 23 p).

40. "Ground Water and Wells," (St. Paul: Johnson Div., UOP, 1975, 440 p).

41. Lehr, J. H., Pettyjohn, W. A., Bennett, M. S., Hanson, J. R. and L. E. Sturtz. "A Manual of Laws Regulations, and Institutions for Control of Ground Water Pollution," (Washington: EPA 440/9-76-006, June 1976, 416 p).

42. Lindorff, D. E. and K. Cartwright. "Groundwater Contamination: Problems and Remedial Actions," (Urbana: Illinois State Geological Survey, Environmental Geology Notes No. 81, May 1977, 30 p).

43. Lohman, S. W. "Ground-Water Hydraulics, (Washington: Government Printing Office, U.S. Geological Survey Professional Paper 708, 1972, 70 p).

44. MacIver, B. N. and G. P. Hale. "Laboratory Soils Testing," (Washington: Dept. of Army, EM 1110-2-1906, Nov. 1970).

45. Malmberg, K. B. "EPA Visible Emission Inspection Procedures," (Washington: EPA, Aug. 1975, 68 p).

46. McNabb, J. F., Dunlap, W. J. and J. W. Keeley. "Nutrient, Bacterial, and Virus Control as Related to Groundwater Contamination," (Ada, Okla.: EPA 600/8-77-010, July 1977, 18 p).

47. Meinzer, O. C. "Kinds of Rocks and their Water-Bearing Properties," in Occurrence of Ground Waters in the United States, (Washington: U.S. Government Printing Office, 1923, pp. 102-148).

48. Miller, D. W. Groundwater Monitoring Components, (Syossett, NY: Geraghty and Miller, Inc., 1979, 7 p).

49. Mooij, H. and F. A. Rovers. "Recommended Groundwater and Soil Sampling Procedures," (Ottawa: Environmental Conservation Directorate, Report EPS-4-EC-76-7, June 1976, 35 p).

50. "A Handbook of Radioactivity Measurements Procedures," (Washington: National Council on Radiation Protection and Measurements, NCRP Report No. 58, Nov. 1978, 506 p).

51. "Basic Radiation Protection Criteria," (Washington: National Council on Radiation Protection and Measurements, NCRP Report No. 39, Nov. 1978, 135 p).

52. "Instrumentation and Monitoring Methods for Radiation Protection," (Washington: National Council on Radiation Protection and Measurements, NCRP Report No. 57, May 1978, 177 p).

53. "Enforcement Considerations for Evaluations of Uncontrolled Hazardous Waste Disposal Sites by Contractors," (unpub.), (Denver: National Enforcement Investigations Center, EPA, April 1980).

54. "NEIC Policies and Procedures Manual," (Denver: National Enforcement Investigations Center, EPA 330/9-78-001, May 1978, 54 p).

55. "NEIC Safety Manual," (Denver: National Enforcement Investigations Center, EPA 330/9-74-002-B, Feb. 1977, 125 p).

56. "Safety Manual for Hazardous Waste Site Investigations," (unpub.), (Denver: National Enforcement Investigations Center, EPA, Sept. 1977).

57. "Occupational Diseases, A Guide to Their Recognition," (Washington: National Institute for Occupational Safety and Health, U.S. Government Printing Office, June 1977, 608 p).

58. "Registry of Toxic Effects of Chemical Substances," (updated yearly), (Washington: National Institute for Occupational Safety and Health, U.S. Government Printing Office).

59. "Pocket Guide to Chemical Hazards," (Washington: National Institute for Occupational Safety and Health, GPO 760-553, June 1979, 191 p).

60. "Available Information Materials on Solid Waste Management, Total Listing, 1966-1978," (Washington: Office of Solid Waste, EPA, 1979, 179 p).

61. "The Prevalence of Subsurface Migration of Hazardous Chemical Substances at Selected Industrial Waste Land Disposal Sites," (Washington: Office of Solid Waste, EPA 530/SW-634, Oct. 1977, 513 p).

62. "Manual of Water Well Construction Practices," (Washington: Office of Water Supply, EPA 570/9-75-001, 1975, 156 p).

63. Palmquist, R. and L. V. A. Sendlein. "The Configuration of Contamination Enclaves from Refuse Disposal Sites on Floodplains," (Groundwater: 13:2, 1975, pp. 167-181).

64. Patty, F. A., Fassett, D. W. and D. D. Irish, eds. Industrial Hygiene and Toxicology, Volume II, Toxicology, (NY: Interscience Publishers, 1963, 2377 p).

95

65. Peckham, A. E. and W. G. Belter. "Considerations for Selection and Operation of Radioactive Waste Burial Sites," Second Ground Disposal of Radioactive Wastes Conference, Atomic Energy of Canada Limited, Chalk River, Canada, Sept. 26-29, 1961, (Washington: Nuclear Regulatory Commission, Book 2, Mar. 1962, pp. 428-436).

66. Pettyjohn, W. A. "Monitoring Cyclic Fluctuations in Groundwater Quality," Third National Groundwater Quality Symposium Proceedings, (Ada, Okla.: EPA 600/9-77-014, June 1977, pp. 116-124).

67. Pfannkuch, H. O. and B. A. Labno. "Design and Optimization of Groundwater Monitoring Networks for Pollution Studies," Third National Groundwater Quality Symposium Proceedings, (Ada, Okla.: EPA 600/9-77-014, June 1977, pp. 99-106).

68. Pritchard, J. A. "A Guide to Industrial Respiratory Protection," (Washington: Government Printing Office 017-033-00153-7, June 1976, pp. 66-71).

69. Reinbold, K. A., Hassett, J. J., Means, J. C., and W. L. Banwart. "Adsorption of Energy-Related Organic Pollutants: A Literature Review," (Cincinnati: EPA 600/3-79-086, Aug. 1979, 170 p).

70. Sax, N. I. Dangerous Properties of Industrial Materials, 5th ed., (NY: Van Nostrand Reinhold, 1979, 1258 p).

71. Sisk, S. W. "Recommended Sediment and Sludge Sample Collection Procedures for Priority Pollutant Analysis," (Kansas City: Workshop on Sampling for Pollutant Fate and Risk Assessment Studies, (unpublished manuscript), EPA, 1978, 12 p).

72. Strahler, A. N. Physical Geography, 3rd ed., (NY: Wiley and Sons, 1969, 733 p).

73. Sunshine, I., ed. Handbook of Analytical Toxicology, (Cleveland: Chemical Rubber Co., 1969, 1081 p).

74. Swift, J. J., Hardin, J. M. and H. W. Calley. "Potential Radiological Impact of Airborne Releases and Direct Gamma Radiation to Individuals Living Near Inactive Uranium Mill Tailings Piles," (Washington: EPA 520/1-76-001, Jan. 1976, 44 p).

75. Thompson, M. M. "Maps of America: Cargographic Products of the U.S. Geological Survey and Others," (Washington: U.S. Government Printing Office, 024 001 03145-1, 1979, 265 p).

76. Thornbury, W. D. Principles of Geomorphology, 2nd ed., (NY: Wiley and Sons, 1969, 594 p).

77. Todd, D. K. Ground Water Hydrology, (NY: Wiley and Sons, 1966, 336 p).

78. Todd, D. K., ed. The Water Encyclopedia, (Port Washington, New York: Water Information Center, 1970, 559 p).

79. Tolman, A. G., Bullestero, A. P., Beck, W. W. and G. H. Emrich. "Guidance Manual for Minimizing Pollution from Waste Disposal Sites," (Cincinnati: EPA 600/2-78-142, Aug. 1978, 83 p).

80. "Chemical and Physical Effects of Municipal Landfills on Underlying Soils and Groundwater," (Cincinnati: U.S. Army Engineers Waterways Experiment Station, EPA 600/2-78-096, May 1978, 140 p).

81. Waldbott, G. L. Health Effects of Environmental Pollutants, 2nd ed., (St. Louis: C. V. Mosby Co., 1978, 350 p).

82. Earth Manual, (Washington: Water and Power Resources Serv., U.S. Government Printing Office, 1980, 810 p).

83. Welch, P. S. Limnology, 2nd ed., (NY: McGraw-Hill, 1952, pp. 33-91).

84. Windholz, M., ed. The Merck Index, 9th ed., (Rahway, NJ: Merck and Company, 1976, 1313 p).

5

NIOSH AIR MONITORING AT HAZARDOUS WASTE SITES

Richard J. Costello, PE, CIH
Diplomate American Academy of Environmental Engineers
Division of Surveillance, Hazard Evaluations and Field Studies
National Institute for Occupational Safety and Health
Cincinnati, Ohio 45226

This chapter discusses strategies for assessing inhalation exposure to chemicals at hazardous waste sites, the available methods for measuring exposures, and some of the limitations of the instruments and methods used. The information presented is an outgrowth of several studies conducted by the National Institute for Occupational Safety and Health (NIOSH) during the past three years at landfills, operational hazardous waste treatment facilities, and CERCLA (Comprehensive Environmental Response, Compensation, and Liability Act) sites, which are commonly called "Superfund" sites. Measuring worker exposure to chemicals, especially exposure by inhalation, is one of the quantitative steps used in developing rational health and safety programs. Measurements are especially needed to assess the health consequences of exposure to a plethora of toxic wastes. Data, especially data about exposure, are necessary before adequate protective measures can be selected.

BACKGROUND

Since the passage of the CERCLA Act, there has been increased concern about the health and safety of workers who clean up hazardous waste sites and who respond to chemical spills. It also has been apparent that the air monitoring technology needed for measurement of occupational exposures at hazardous waste sites has been either absent or is not well developed. This problem was highlighted in late 1980 by the Surgeon General [1], who advised the Senate Committee on Environment and Public Works Committee:

> We agree that exposure to toxic chemicals is a major public health concern; however, because of the enormity of the problem and current limitations in our knowledge base, we are unable to provide you with as complete an assessment as we would wish. . . [We] are currently in the early stages of a health problem that may take years to assess fully.

The Surgeon General's Report emphasized the lack of scientific methods for evaluating the health effects of chemical waste dumps on human health, including the lack of adequate tests to evaluate the effects of chemicals in human populations; the lack of information for identifying complex mixtures of chemicals; the lack of data and experience in testing mixtures of chemicals for potential health effects; and the lack of ways for dealing with the time lag between chemical exposure and the manifestation of health effects. The report suggested that epidemiological studies were of limited use for assessing potential health effects of toxic waste because of the small number of people affected at each site; the lack of exposure data; and the lack of common, clinical signs indicative of a specific disease.

Nonetheless, in December 1980, the CERCLA Act [2] provided that,

> The President, acting through the Administrator of the Environmental Protection Agency, the Secretary of Transportation, the Administrator of the Occupational Safety and Health Administration, and the Director of the National Institute for Occupational Safety and Health shall study and, not later than two years after the enactment of this Act, shall modify the national contingency plan to provide for the protection of the health and safety of employees involved in response actions.

The Need for Contaminant Identification and Quantification

Reliable measurements of worker exposure to airborne contaminants at hazardous waste sites are necessary to select appropriate protective equipment, based on the potential health effects of the exposure; to differentiate between areas where protection is needed and areas where it is not needed; and to demonstrate repetitive exposures so that medical monitoring can be initiated, if needed. Industrial hygiene has two principal techniques for quantification of airborne contaminants. One method employs direct-reading instruments which, when properly calibrated, respond quantitatively to a limited number of chemical substances. The second technique, which requires laboratory analysis of the samples, uses grab, filter, sorbent, or wet contaminant collection methods.

But no matter how the samples are collected, accurate identification of the contaminants must precede accurate quantification if optimal, and in many instances merely useful, results are to be obtained. This is especially true when the air samples may contain complex mixtures of contaminants. There can be a large number of potential airborne contaminants coexisting in the workplace atmosphere even at small sites (see Tables I and II). Once the identity of the contaminants is known, appropriate direct-reading instruments can be selected (if available) for immediate field measurements. If samples are collected for laboratory analysis, identification of contaminants is still necessary to develop optimum sampling and analytical protocols. If the identification of the major airborne contaminants is incomplete, for example if only volatile organics are identified when particulate or particulate-bound contaminants are the major exposure, then any assessment of health and safety conditions based solely on the volatile data will be incomplete. Incomplete data can easily result in inadequate or inappropriate worker protection. Reference 17 describes a study which illustrates this point.

Regulatory Guidelines

The difficulties associated with the identification of the components of complex mixtures of airborne chemicals and

Table I.   Estimated Volume of Chemicals [15]
Triangle Chemical Site
Bridge City, TX

Number of Drums	Contents (gal)	Volume
260	Solvents	10,400
60	Acids	3,600
90	Bases	3,600
175	Alcohols	7,000
85	Ether	3,400
250	Empty	0
		28,000

Table II. Estimated Frequency of Chemicals [15]
Triangle Chemical Site
Bridge City, TX

Category	Compounds	Relative Abundance
Solvents...	Dichlorobenzene Toluene Trichlorethylene Xylenes Methylethylketone Orthodichlorobenzene	37%
Acids......	Cresylic Acid Hydrochloric Acid Hydrofluoric Acid Surfactant (pH 1) Nitric Acid Phosphoric Acid	13%
Bases......	Ammonium Hydroxide Caustic (pH 12) Diethanolamine Methylethylamine	13%
Alcohols...	Butanol p—Decyl—phenol Ethanol Glycols Rust tem 200 Methanol p—Nonylphenol	25%
Ethers.....	Ether	12%
		100%

the limitations of available instruments for rapidly monitoring a mixture of unknown airborne contaminants at hazardous waste sites became apparent shortly after the passage of CERCLA, when the U.S. Environmental Protection Agency developed a "Draft Interim Site Safety SOP (Standard Operating Procedure)" [3] for worker protection in multiple toxic chemical environments. It suggested air sampling with portable instruments and use of "total organic vapor" readings obtained with flame ionization and photoionization detectors for the selection of levels of respiratory and skin protection. A 1982 version of this document stated:

> When the presence of organic
> vapors/gases are unknown, instruments
> such as the photoionizer . . . and/or
> the portable gas chromatograph . . .,
> operated in the total readout mode
> should be used to detect organic
> vapors. Until specific constituents can
> be identified, the readout of the
> instruments indicate total airborne
> substances to which the instrument
> responds. . . .
> These gross measurements can be used on
> a preliminary basis to . . . determine
> levels of personal protection and select
> candidate areas for more thorough
> qualitative and quantitative studies.

The levels of protection prescribed emphasized the use of self-contained breathing apparatus and (in the initial versions of the Draft SOP) the extensive use of fully encapsulating suits. Given the undesirable health effects of heat stress and the 25–40% manpower efficiency penalties associated with wearing extensive personal protective equipment [4], this approach was apparently made necessary by the lack of timely information about contaminant identity and concentration which could then be used for the selection of protective equipment.

The "Draft SOP" was proposed as national policy in the Federal Register of March 12, 1982 [5]; however, after public comments, the formal requirement that On-Scene Coordinators (OSCs) use the Interim Standard Operating Safety Procedures was deleted [6]. The current policy is:

> [The] OSC or responsible official must
> conform to applicable OSHA requirements
> and other guidance. All private
> contractors who are working at the scene

of a release must conform to applicable
provisions of the Occupational Safety
and Health Act and any other
requirements deemed necessary by the
lead agency.

The identification of contaminants and quantification
of exposure is the prerequisite for developing
scientifically based worker protection measures. There is
little quantitative occupational exposure information
available from hazardous waste sites. In the three years
since the passage of the CERCLA Act, selection of
respiratory protective equipment at hazardous waste sites
has often been made in the absence of accurate measures of
the nature or extent of the inhalation hazards. There is no
established exposure/effect relationship for the health
consequences of inhaling total organic vapors measured with
a flame ionization detector or photoionization device. This
method, which is frequently used to select both respiratory
and skin protection at hazardous waste sites, is not based
on established principles the exposure/effect relationship.
Reliable measurements of airborne contaminants are
necessary: 1) to select appropriate protective equipment,
based on the potential health effects of the exposure; 2) to
differentiate between areas where protection is needed and
areas where it is not needed; and 3) to demonstrate
repetitive exposures so that medical monitoring can be
initiated, if needed. These data should also be published
to assist other health and safety practitioners. There is
no device for measuring the penetration of liquid chemicals
through the intact skin outside of the laboratory. Skin
exposure cannot be assessed by measuring levels of airborne
"total vapors" [7,8].

The Types of Occupational Exposure

There are two generally recognized kinds of adverse
health effects induced by chemical exposures at work. Acute
effects result in undesirable and irreversible health
changes in short periods of time (in the order of magnitude
of a few seconds or minutes). Chronic effects reflect the
cumulative bodily damage resulting from repetitive exposures
that do not produce immediately irreversible consequences.
The exposures occur again and again during long periods of
time (in the order of magnitude of years). The techniques
for measuring the lower levels of exposure that typically
produce chronic health effects frequently differ from those

used to measure exposures which may result in acute effects. Because immediate decisions concerning protection are usually required, direct-reading instruments are ordinarily used to evaluate exposures likely to result in acute effects. Full-shift personal air samples, analyzed in an off-site laboratory, are ordinarily used to assess exposures believed to be below the immediately dangerous to life and health (IDLH) range. Long-term or full-shift direct-reading monitors are available for a few substances, but oftentimes lack the necessary sensitivity or accuracy. Health and safety plans that require air monitoring must recognize the difference between these two kinds of health effects and the techniques used to measure them. Common to both techniques is accurate identification of contaminants to be measured. This topic will be discussed in the context of conditions encountered at hazardous waste sites.

The Variability of Hazardous Waste Site Exposure with Location, Operation, and Time

In complex, multi-substance environments where chronic exposures are suspected, production line or unit operation oriented practice suggests that screening surveys employing samples qualitatively analyzed in the laboratory precede quantitative exposure measurement. This may involve two site visits separated, at a minimum, by the analytical turnaround time of the screening samples. The turnaround time can be as little as 24-48 hours if priority analysis resources are available, but is usually several weeks or months. In the alternative, a subset of the samples collected during a single visit can be analyzed qualitatively and the balance of the samples from the same visit can be analyzed for specific compounds based on the qualitative findings.

An important assumption implicit in the multi-visit monitoring technique is the steady-state composition and concentration of contaminants. Common sense suggests that, in the absence of active handling of contaminated materials, worker exposures would be stable or would change only slowly above undisturbed hazardous waste sites; however, steady-state conditions are not invariably the rule. Dusts, including both finely divided hazardous solids and other hazardous materials coated onto soil particles, are highly sensitive to a number of factors that can vary significantly with location and time, especially soil moisture content. Vapor emissions can be produced by mechanical displacement of saturated vapors, which can produce relatively high short-term concentrations; direct evaportations, which can can produce moderate short-term concentrations; or diffusion mechanisms, which can can produce relatively low short-term concentrations. Direct evaporation and diffusion can also be important long-term phenomena since they may continue for long periods of time and involve large areas. Again, water can essentially cap or plug vapor emission routes and greatly reduce airborne emissions [9]. Seasonal temperature changes may influence conditions, since a 10°C increase in soil temperature can more than double the vapor pressure of PCBs and an increase of windspeed from 3.7 to 6.9 m/s at 20° C can double the PCB vapor concentration near a free liquid surface [4].

In addition to the environmentally induced variability of exposure, once a remedial action starts chemically contaminated materials are actively handled, and it is reasonable to suspect that the composition and concentration of chemical substances in a worker's breathing zone varies substantially from point to point on the site. This would be due in part to the different processes or types of operation at various site locations and in part to the variable composition of the waste from point to point. Differences in exposure with site location have been demonstrated for individuals who work full shifts in and around trenches in contaminated soils and workers who handle contaminated materials directly [10,11].

The multi-visit sampling technique measures location- and operation-specific exposures at the time of the study. Lengthy laboratory delays mean that exposure information is not available for use in protecting workers at the time exposures are taking place. This is a serious problem because hazardous waste site cleanup does not resemble a repetitive production line operation, but rather it resembles the constantly changing stages of completion characteristic of construction projects. Before the "results" are available from the laboratory analysis, the cleanup may well progress to different site locations where different contaminants or different mitigation operations are performed. Conversely, if a one-site visit strategy is used and a limited number of samples are screened qualitatively and the balance analyzed on the basis of these results, then differences in exposure at various parts of the site may be masked because of contaminant and concentration differences at various site locations.

Because screening samples are collected in a short time span (at most a few days), they are similar to a snapshot of site exposure conditions. The information needed for occupational health purposes is a series of samples collected sequentially in time, a technique similar to the individual frames of film exposed in sequence in a movie camera. These results can be used to differentiate between repetitive exposures and isolated or one-shot exposures. There are three ways to get this information: with direct-reading instruments, with samples analyzed in a laboratory, or with a combination of the two. In any measurement problem, however, the first consideration is what to measure.

What to Measure: Selecting Contaminants for Monitoring

Long-Term Effects

Much of this chapter will focus on techniques to evaluate 8-hour inhalation exposures during hazardous waste removal actions. These are preplanned sampling operations. Because background information is available to identify prospective contaminants, air monitoring at cleanups is not subject to the uncertainties inherent in emergency response actions. For planned sampling visits to evaluate suspected chronic hazards, a useful strategy is to first examine the background data available to help select the site contaminants to be measured. Two factors, the probable amount of a contaminant or class of contaminants (e.g., pesticides, polynuclear aromatic hydrocarbons (PNAs), or polychlorinated biphenyls (PCBs) on the site and the contaminant toxicity, are considered. The substances or classes of substances are ranked and as many analytes selected as can be accommodated within laboratory workload and financial restraints. This approach favors selection of materials present in large quantities. It discriminates against substances present in relatively small quantities, unless the health hazard of even infrequent exposure is deemed overriding. Finally, it assumes that all contaminants, whether solids, liquids, gases or vapors, are equally likely to be released. Broad-spectrum sampling is favored.

The most frequent source of site contaminant information is the results of drum, barrel, or bulk composite samples. Since the volume of waste in various chemical classes is usually estimated to assess the cost of removal or treatment, these estimates can be used to choose substances to be measured. At this stage, no judgement is made about the probability of release; deteriorated barrels are considered as likely to rupture as new barrels, and the release of contaminants from open ponds is considered as likely as the release of contaminants from emptying storage tanks. Simply restated, the process of choosing the substances to be measured for chronic exposures assumes that the substances present on the site in the largest quantities are those most likely to be frequently handled and most likely to be repetitively released into the workplace air. Substances that are not themselves a site contaminant, but that can be generated by the chemical reaction of site contaminants, must also be considered. Standard listing of hazardous chemical reactions are consulted [12], especially if the potential reactants are located near to each other.

The physical state of a contaminant must also be considered. Many contaminants such as PCBs, PNAs, or pesticides should be measured as both particulate-bound contaminants and as vapors. The volatile component is collected on a solid adsorbent and the non-volatile component is collected on a filter. More than two dozen methods have been developed by NIOSH [13,14] which make use of a dual media sampling system to efficiently collect both the particulate and vapor portion of a single contaminant, e.g., MOCA. Measurement techniques that analyze only the vapor phase of a substance may fail to identify significant exposure to toxic contaminants if the substance is primarily particulate or particulate bound. Some of the sampling media and analytical techniques that have been used during NIOSH studies at hazardous waste sites are listed in Table III. If all workers are exposed to all contaminants, the logistics of sample collect become difficult.

Acute Effects

In emergencies, however, there are little background data, although acute exposures immediately dangerous to life and health (IDLH) are suspected. IDLH exposures require immediate monitoring to determine the degree of protection needed or the extent of withdrawal necessary to protect human life. At hazardous waste sites, unanticipated and undefined exposures are usually handled as if they are potentially IDLH (see Figure 1). Immediately dangerous conditions which can be monitored with traditional industrial hygiene instruments include oxygen deficiency, explosive vapors, and acutely toxic concentrations of known chemicals. It is important that the identity of the chemical contaminant be ascertained so that appropriate measuring instruments can be selected. In the absence of contaminant identification, withdrawal or use of air-supplied respirators is necessary.

Table III. Sample Collection and Analytical Methods

Substance	Collection Device	Analytical Method	Typical Limit of Detection (μg)
Anions:	Prewashed Silica Gel Tube	Ion Chroma- tography	
Chloride			5
Nitrate			10
Bromide			10
Fluoride			5
Sulfate			10
Phosphate			20
Aliphatic Amines	Silica Gel	GC/NPD	10
Asbestos	AA Filter	PCM	4500*
Metals	AA Filter	ICP-AES	0.5
Organics I	Charcoal Tube	GC/MS	10
Nitrosamines	Thermosorb/N	GC-TEA	0.01
Particle Size Distribution	Personal Cascade Impactor	Gravi- metric	
PCBs	GF Filter and Florisil Tube	GC/HECD/ECD	0.05
Pesticides	13mm GF and Chromosorb 102 Tube	GC/MS	0.05

NOTES:
1. ICP-AES means inductively coupled plasma atomic emission spectrometry; GC/MS means gas chromatography and mass spectrometry; IC means ion chromatography; NPD means nitrogen/phosphorus detector; FID means flame ionization detector; HECD means Hall Electrolytic Conductivity Detector; TEA means thermoelectron detector; PCM means phase contrast microscopy.
2. GF means that a glass fiber filter was used for sample collection.
3. *Units in fibers per filter.

Figure 1. Acute exposure at a hazardous waste site.

Where to Measure: Detail of the Industrial Hygiene Air
Monitoring Strategy

Long-Term Effects

A useful monitoring protocol for characterizing
repetitive exposures at hazardous waste sites [10,15,16]
hypothesizes three more or less concentric exposure areas
(Figure 2). In the center, Area "A", the highest exposures
to the most substances are anticipated. Waste materials are
actively handled by workers. Drum ruptures or conditions
conducive to the unplanned release of waste chemicals are
most likely. Drum moving and drum opening are typical
operations. Personnel utilize the highest levels of
personal protection required on the site.
Area "B" is the concentric circle that surrounds Area
"A". In Area "B", materials are handled indirectly.
Unplanned releases of chemicals are less likely and
personnel may wear a lower level of personal protection than
those involved with direct material handling. Drum
restaging and contamination reduction are typical operations
in the area. Area "C" is the site clean zone. In Area "C",
toxic substances are not handled and personal protective
equipment is not required. The factors used to
differentiate the areas are the amount of active material
handling and the intimacy of worker contact with the
materials. If workers manipulate drums with their hands and
the drums frequently rupture during this handling, the area
would be designated "A". If drums are stacked on pallets
and occasionally rupture when moved with forklifts, the area
would be designated "B".
The pattern of concentric exposure zones also suggests
that personal air samples collected in the breathing zone of
workers in the "A" Area give an estimate of the maximum
credible site exposure. Fixed location monitoring at the
"fenceline" or perimeter where protective equipment is no
longer required serves as a measure of contaminant migration
away from the site and a measure of the integrity of the
site clean zones. Since the fixed-location samples may
reflect exposures primarily either upwind or downwind from
the site, wind direction and velocity data are needed to
interpret these sample results.
The 1980 Surgeon General's Report noted that very few
individuals were exposed to hazardous wastes originating at
any particular site, and this is true whether occupational
or community exposures are considered. Two to three dozen
workers is a "large" waste site worker population. It is

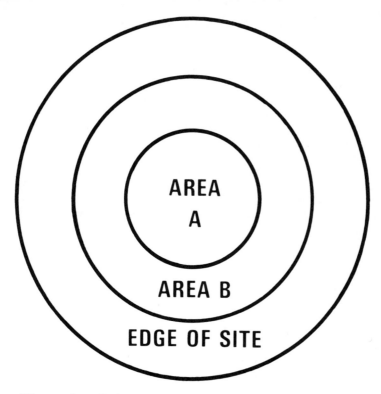

Figure 2.   Industrial hygiene sampling strategy.

not feasible to collect samples on a single worker using a
variety of sampling media during a single day. Because of
the strenuous physical activity and the amount of personal
protective equipment usually worn, hazardous waste workers
cannot usually wear more than one monitoring device.
Because it is not usually possible to draw air through
different sampling media using a single portable
battery-operated pump, it requires several days to measure
the exposure of a specific individual using each of the
media [15,16]. One way to overcome this problem is to
collect multiple area samples on pieces of heavy equipment.
While technically not a personal sample, heavy equipment
operators remain at the vehicle operators position
throughout most of the workshift and usually are employed in
the active materials handling part of the site. Samples
collected very close to the breathing zone are reasonably
representative of personal exposure and these multi-media
samples yield as much information as several personal
samples [17].

Acute Effects

     Chemical exposures that may result in acute health
effects are most likely to be found in confined spaces.
Sumps, silos, cargo holds, storage tanks, and mine shafts
are notorious examples. Because toxic materials emitted
into the atmosphere tend to be transported away from the
source and simultaneously diluted, acutely dangerous
conditions are not likely to persist in open spaces for
extended periods of time unless there is a very large (and
hence readily identifiable) source, such as an overturned
tankcar. The account of the air contaminant measurements
made at the Mississauga train derailment illustrates this
point [18,19]. Absent a large persistent source,
immediately dangerous inhalation hazards in the open
atmosphere are likely to last for only brief periods, such
as in the immediate vicinity of a leaking gas cylinder;
however, they can persist in confined spaces over long
periods of time.

Long-Term Effects

For research purposes, measurements to estimate the mean exposure and its variance are desirable. In such studies, all of the workers on a site may be monitored during several consecutive days. For routine operations, however, the NIOSH Occupational Exposure Sampling Strategy Manual [20] suggests selective monitoring of high-risk workers; specifically, monitoring those workers who are closest to the source of contaminant generation. The rationale for this procedure is that the possibility of a significant exposure varies directly with distance from the source. If the high-risk workers are not exposed significantly, then monitoring high-risk workers first conserves resources that would otherwise be necessary to monitor workers further removed from the contaminant source.

In multi-substance environments, where simultaneous exposures are anticipated and where the contaminants must be collected on a number of different media, it may be necessary to sample more than one worker in each operation. Repetitive sampling of more than one worker in each operation is recommended. Sampling should continue throughout the cleanup, unless it can be demonstrated that the site contaminants are homogeneously distributed and that consistent exposures would be anticipated.

Repetitive sampling might be discontinued if sampling in maximally contaminated areas has produced negative air sampling results. For example, if soil and ground water samples have been collected in consistent locations and strata, the contaminant data could be transformed to consistent units and the numbers summed. A plot of these calculated values would then suggest site areas where more intensive exposure monitoring efforts may be required (Figure 3). The procedure could be repeated for total contaminants or for subsets of contaminants. Of course, this technique does not predict the magnitude of the potential exposure, only the locations where relatively higher exposures may occur. Air sampling must still be conducted, but monitoring efforts would be focused on areas of highest suspected exposures.

CONTAMINATION OVERVIEW

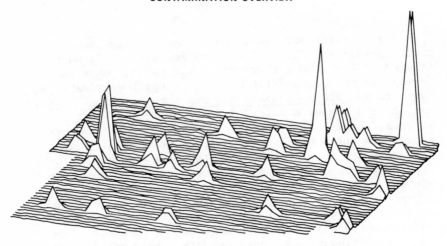

PEAKS REPRESENT TOTAL CONCENTRATION

Figure 3. Summed soil and groundwater data used to identify more highly contaminated site areas.

Acute Effects

Since exposures likely to induce acute health effects are incident specific, it is not possible to identify candidates for monitoring in advance.

When to measure. The use of models for range-finding estimates of potential worker exposure to waste site dusts and vapors has been suggested; however, this technique does not appear to be practical in the microenvironments of occupational exposure. A recent study concluded that, even under the best of conditions, the accuracy of existing models is highly uncertain [9].

Air sampling to identify immediately dangerous conditions should be made as soon after site identification as possible. These measurements should be followed by characterization of contaminants at the site perimeter as an index of the migration of contaminants away from the undisturbed site. Since occupational exposures are linked closely with active material handling, personal air sampling should not be necessary until site mitigation has begun. A reasonable air sampling procedure would involve an initial site characterization using a variety of sampling media to identify the major classes of contaminants that actually become airborne and the levels of airborne contamination. This information would be useful in the selection of levels of respiratory protection for routine or day-to-day site operations. This initial characterization should be followed by daily sampling of a small number of "high-risk" workers for the contaminants most frequently present and in the highest concentrations. It may be necessary to periodically repeat a broad spectrum sampling scheme, especially when work is begun on a markedly different portion of the site where markedly different contaminants or when a markedly different type of site operation (such as barrel opening as opposed to exploratory well drilling) is initiated. Repeated broad spectrum scans serve as a control for variation of contaminant types and emission rates with site location. Limited sampling of "high-risk" workers should continue until the next broad spectrum scan.

How to measure. Industrial hygiene measurements at hazardous waste sites are made with the same equipment as industrial hygiene measurements are made in any other environment. Laboratory samples are analyzed in the same way as samples collected in general industry. Development of sampling protocols and interpretation of the measurements made at hazardous waste sites are more difficult because many complicating factors are present. The factors peculiar to hazardous waste sites include:

117

* The large number of (often unidentified) bulk
  chemicals present and the complexity of the
  mixtures. Industrial chemicals are usually pure.
  Wastes are not pure nor were they manufactured to
  conform with any specification.
* The variability of exposure with location without a
  corresponding change of process or unit operation.
  Traditionally, industrial hygienists have
  catagorized exposures in terms of their location in
  the production of a product. For example,
  production line station #1 might be associated with
  particulate heavy metal exposures at levels usually
  associated with chronic health effects, while
  station #2 is associated with acutely toxic
  vapors. Workers from station #1 would not
  ordinarily be assigned to station #2, and so the
  exposure populations are mutually exclusive. At
  waste sites, workers may be exposed to both heavy
  metals and toxic vapors as they move from point to
  point on the site. The exposure populations are
  not mutually exclusive.
* Hazardous waste cleanup usually occurs in an
  outdoor environment. In-plant exposures are
  ordinarily confined to the immediate vicinity of
  their source. Given a localized source,
  contaminants are not usually transported from one
  part of the plant to another. In outdoor
  environments, airborne contaminants are readily
  transported from one part of the site to another.
  Workers are not only exposed to the materials which
  they handle directly, they are also exposed to
  materials handled elsewhere on the site.
* The quality of the waste containers is poor.
  Commercial products are packaged in containers that
  do not leak under ordinary circumstances. However,
  deteriorated containers at waste sites routinely
  rupture and release their contents.
* The lack of containment vessel integrity and the
  incomplete identification of contaminants increase
  the possibility that waste mixing will occur,
  resulting in dangerous reaction products.
* Workers may receive multiple exposures. Factors 1,
  3, and 5 result in complex mixtures of airborne
  contaminants.

The result of these factors is that real-time
identification of contaminants is important at hazardous
waste sites. NIOSH has developed two approaches for near
real-time contaminant identification: the SDRITS and the
Mobile Operational Base.

118

Portable, direct-reading instruments are useful for characterizing exposures that may produce acute health effects. Unlike conventional samples that must be analyzed in a laboratory, exposures detected by direct-reading instruments are measured in real time and are available for immediate decision making. However, the limitations of the instruments used must be recognized and considered when interpreting the results. Most portable instruments are designed for measuring specific materials in a single phase or physical state. There is no one device that can rapidly monitor multi-contaminant atmospheres for all of the solids, vapors, or gases that can produce acute adverse health effects. Some instruments are hampered by cross-sensitivites; that is, devices intended to measure substance "A" also respond to substance "B", and the responses are indistinguishable. In uncharacterized atmospheres, an instrument response to substance "B" can be erroneously interpreted as an indication of the presence of "A". If photoionization or flame ionization instruments are used in the "total vapor" mode, particulates will not be detected. Meter response would be a function of the kinds of materials present and their instrument response factors, as well as the concentration present, and individual substances will not be quantitated [21,22].

Despite the limitations of available direct-reading instruments, the rational selection of protective equipment and even the information to determine whether protective equipment is needed depends on knowledge of the identity and concentration of contaminants. In uncharacterized environments, a procedure for qualitative identification of contaminants can expedite selection of appropriate direct-reading instruments, assist in establishing proper instrument conditions for monitoring, and provide information for selection of sampling media to be analyzed later in the laboratory. An experimental device, the SDRITS, or Simultaneous Direct-Reading Indicator Tube System [23], can qualitatively screen for broad classes of contaminants (Figure 4). Halogenated hydrocarbons, aromatic hydrocarbons, amines or reactive nitrogen materials, acid reacting substances, ketones, and alcohols can be distinguished using previously published protocols [24,25]. The SDRITS is an application of existing direct-reading indicator tube technology. Contaminants of local interest, such as particulate cyanides, or gases, such as hydrogen sulfide, can also be determined. Since the SDRITS uses battery-operated pumps to draw air through 10 indicator tubes simultaneously, it is much faster than traditional direct-reading indicator tube measurements.

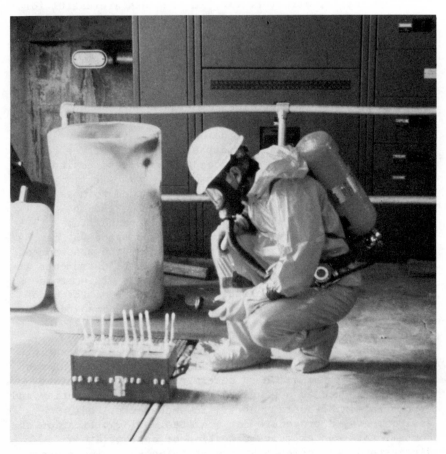

Figure 4. The SDRITS.

Lengthy laboratory delays for sample analysis mean that exposure information is not available for use in protecting workers at the time exposures are taking place. This is a serious problem because, as stated earlier, hazardous waste site cleanups do not resemble repetitive production line operations, but rather resemble the constantly changing stages of completion characteristic of construction projects. One way to decrease the lag time between sample collection and sample analysis is on-site sample evaluation. To demonstrate this concept, NIOSH has developed a mobile operational base for use at hazardous waste sites.

The mobile base provides working space suitably equipped for scientific equipment, a resource that is generally not available at hazardous waste sites. It provides storage space, countertop staging areas for industrial hygiene equipment, and facilities for SCBA recharging. It also houses analytical instruments capable of classifying contaminants by a variety of techniques.

While the mobile base does not duplicate the capabilities of a well-equipped laboratory, it does accommodate facilities for screening and range-finding procedures conducted in near real time. If indicated, samples screened in the mobile base can be reanalyzed in sophisticated fixed-base laboratories. Typical instruments available in the mobile base include gas chromatographs, ion chromatographs, and X-ray fluorescence spectrometers. When not in use in the mobile base, these devices can be removed and are available for use in fixed facilities. While only a few of the field samples collected are analyzed on site, the results provide immediate range-finding estimates of the concentration of airborne contaminants which is useful for assessing the general scheme of personal protection. The data developed can be used to guide the fixed-base laboratory analysts and to modify field sampling procedures, if necessary.

Future plans for the mobile base include field tests of advanced portable X-ray fluoresence analyzers [26,27], which offer real-time field elemental analysis of particulates and are non-destructive. Additional experiments will test the application of Fourier Transform Infrared Spectrometry [28,29], which has shown promise for near real-time analysis of both ambient air and bulk waste samples.

Other Industrial Hygiene Considerations

Personal samples are collected in the breathing zone and outside the facepiece, if respiratory protective equipment was being worn. These samples represent the potential inhalation exposure of workers who are not wearing respiratory protection. It is best to use flow-rate-controlled pumps. Observing and adjusting pumps while wearing gloves, respirators, and other kinds of personal protective equipment is difficult. The pumps should be protected with disposable coverings, such as small plastic bags, especially if the pumps are subjected to procedures such as being washed down with a fire hose.

FIELD STUDY RESULTS

During the last three years, NIOSH has collected 500 air samples at hazardous waste sites. About half of these were collected at remedial actions where drums were being actively handled [15,16]. The other half of the samples were collected at a municipal landfill that accepted limited types of hazardous waste [30] and at a hazardous waste treatment facility that accepted virtually all categories of waste [17].

Of the 500 samples, no contaminant exceeded 10% of the OSHA 8-hour permissible time-weighted average concentration. The substances analyzed included organic vapors, heavy metals, PCBs, pesticides, cyanides, acids, and bases. The metals scan typically included 30 elements from aluminum to zinc. There was evidence that particulate-bound PNA materials were responsible for eye and respiratory irritation among workers and possibly among community residents near one of the sites [17].

Table IV shows toluene levels at six sites. The OSHA 8-hour TWA for toluene is 750,000 $\mu g/m^3$. The first four results are only a small fraction of the regulatory standard. Exposures were higher near trenches [31,32], although the concentrations were still low. Trenches impede dilution and dispersion of contaminants. Both the Picillo Farm and Love Canal data (Table IV) support this conclusion.

Table IV.   Toluene Results

	Mean $\mu g/m^3$	Standard Deviation	Max $\mu g/m^3$	Number of Observations
CHEMICAL CONTROL [16] Elizabeth, NJ	900	1.9	6000	12
ROLLINS ENVIRONMENTAL [17] Baton Rouge, LA	410	2.6	2600	32
TRIANGLE CHEMICAL [15] Bridge City, TX	170	4.4	1700	14
FOUNTAIN AVENUE LANDFILL [30] Brooklyn, NY	159	2.1	1600	37
PICILLO FARM [31] Coventry, RI	14,800	--	26,600	4
LOVE CANAL [32] Niagara Falls, NY	5800	--	49,000	69

	OSHA STANDARD	750,000
	NIOSH	375,000

123

SUMMARY

The inhalation exposure data that NIOSH has gathered at
hazardous waste sites during the last three years are well
below recognized occupational standards. This suggests that
the respiratory protection selection criteria in use at
hazardous waste sites (total vapor concentration measured
with a photoionization or flame ionization detector) should
be reconsidered. Furthermore, development of real-time
monitoring devices and techniques is needed if accurate
measures of exposure are to be made routinely. The current
use of total vapor response of organic vapor analyzer or
photoionization instruments for selection of repiratory and
skin protective equipment is not based on accepted health
protection concepts. NIOSH studies have demonstrated that
contaminants at hazardous waste sites can be identified and
quantified on-site using techniques such as the SDRITS and
the mobile operational base.

As the third year since the passage of CERCLA ends, the
comment of the 1980 Surgeon General's Report is still
applicable:

> We agree that exposure to toxic
> chemicals is a major public health
> concern; however, because of the
> enormity of the problem and current
> limitations in our knowledge base, we
> are unable to provide you with as
> complete an assessment as we would wish.
> . . .

REFERENCES

1. "Health Effects of Toxic Pollution: A Report from the Surgeon General and a Brief Review of Selected Environmental Contamination Incidents with a Potential for Health Effects." The Surgeon General Department of Health and Human Services and the Congressional Research Service of the Library of Congress, GPO (1980), p.III.

2. CERCLA Section 301(f).

3. "Interim Standard Operating Safety Procedures (revised June 1982)," U.S. Environmental Protection Agency Office of Emergency and Remedial Response Hazardous Response Support Division (1982), p.4.1-4.2

4. Kinman, R. N. and D. L. Nutini. "Production, Migration and Hazards Associated with Toxic and Flammable Gases at Uncontrolled Hazardous Waste Sites" report for U.S. Environmental Protection Agency Municipal Research Laboratory Office of Research and Development, pp.3,4,41,43.

5. Federal Register, 47(49):10994 (March 12, 1982).

6. Federal Register, 47(137):31218, (July 16, 1982).

7. Coletta, G. C. et.al. "Development of Performance Criteria for Protective Clothing Used Against Carcinogenic Liquids," report prepared for the National Institute for Occupational Safety and Health by Arthur D. Little, Inc. Cambridge, Ma, (1978).

8. Lynch, J. R. "Measurement of Worker Exposure," in Patty's Industrial Hygiene and Toxicology, Cralley, L. V., and L. J. Cralley Eds. (New York: John Wiley and Sons, 1979), p.238-239.

9. Stankunas, A. R. "Modeling Approaches for Estimating Airborne Hazardous Waste Concentrations," report prepared for the National Institute for Occupational Safety and Health by TRC Consultants, East Hartford, CT, (1983).

10. Costello, R. J. and M. V. King. "Worker Inhalation Exposure Monitoring During Removal of Hazardous Waste From a Superfund Site" paper presented at the American Industrial Hygiene Association, Philadelphia, PA, May 22, 1983.

11. Costello, R. J., and M. V. King. "Protecting Workers Who Cleanup Hazardous Waste Sites," Amer. Ind. Hyg. Assoc. J. 43:12-17 (1982).

12. Committee on Hazardous Chemical Reactions. "Manual of Hazardous Chemical Reactions," National Fire Protection Association, Boston, (1975).

13. Hill R. H., and J. E. Arnold. "A Personal Air Sampler for Pesticides," Arch. Environ. Contam. Toxicol., 8:621-28 (1979).

14. Taylor, D. G., Ed. NIOSH Manual of Analytical Methods 2nd. ed. (Cincinnati, The National Institute for Occupational Safety and Health, (1977), p. S-236.

15. Costello, R. J. "U.S. Environmental Protection Agency Triangle Chemical Site, Bridge City, Texas," NIOSH, Health Hazard Evaluations Determination Report HETA 83-417-1357, (1983), pp. 6-7.

16. Costello, R. J., and J. Melius. "Technical Assistance Determination Report, Chemical Control, Elizabeth, New Jersey, TA 80-77," The National Institute for Occupational Safety and Health, (1981), pp. 20-22.

17. Costello, R. J., B. Froenberg, and J. Melius. "Health Hazard Evaluation Determination Report, Rollins Environmental Services, Baton Rouge, Louisiana, HE 81-37," The National Institute for Occupational Safety and Health, (1981), pp.25-38 and 40-46.

18. Sciex, Inc. "Mississauga Dereailmen - Ground Level Concentration Measurements of Chlorine and Detection of Other Compounds," report for Ontario Ministry of the Environment (1980).

19. "Mississauga Derailment," Air Resources Branch, Technology Development and Appraisal Section, Monitoring and Instrumentation Development Unit. Ontario Ministry of the Environment (1980).

20. Leidel, N. A., K. A. Busch, and J. R. Lynch. "Occupational Exposure Sampling Strategy Manual," NIOSH (1977), pp.33-34

21. Lauderdale, Jerry F. "Direct Reading Instruments for Analyzing Airborne Gases and Vapors," in Air Sampling Instruments for Evaluation of Atmospheric Contaminants, (Cincinnati, OH: American Conference of Governmental Industrial Hygienists, (1978) p. U-67 and p. U-162

22. Mathamel, M. S. "Hazardous Substance Site Ambient Air Characterization to Evaluate Team Entry Safety." in Proceedings of the Management of Uncontrolled Hazardous Waste Sites 1981, (Silver Spring, MD: Hazardous Material Control Research Institute, (1981), p.281.

23. King, M. V., P. M. Eller and R. J. Costello. "A Qualitative Sampling Device for Use at Hazardous Waste Sites," Amer. Ind. Hyg. Assoc. J. 44:615-618 (1983).

24. Leichnitz, K. "Qualitative Detection of Substances by Means of Draeger Detector Tube Polytest and Draeger

Detector Tube Ethyl Acetate 200/a," Draeger Review, 46:13-21 (1980).

25. Schneider, D. "The Draeger Gas Detection Kit," Draeger Review 46:5-12 (1980).

26. Rhodes, J. R. "Portable Elemental Survey Meter for Air Contaminants," paper presented at the Electron Microscopy and X-Ray Applications to Environmental and Occupational Health Analysis Symposium, Pennsylvania State University, October 15-17, 1980.

27. Baron, P. A., and J. R. Rhodes. "A Portable XRF Analyzer for 25 Elements," paper presented at the Second NIOSH Scientific Symposium, Rockville, MD, October 30, 1979.

28. Herget, W. F. "Measurement of Gaseous Pollutants Using a Mobile Fourier Transform Infrared (FTIR) System," Proceedings of the 1981 International Conference on Fourier Transform Infrared Spectroscopy, The International Society for Optical Engineering, Bellingham, WA (1981)., pp.449-456.

29. Herget, W. F., and J. D. Brasher. "Remote Measurement of Gaseous Pollutant Concentrations Using a Mobile Fourier Transform Interferometer System," Applied Optics 18:3404-3420 (1979).

30. Costello, R. J., M. Singal, and G. Liss. "A Health Hazard Evaluation of Waste Disposal Operators at a Municipal Landfill," paper presented at the American Industrial Hygiene Association Conference, Philadelphia, PA, (May 24, 1983).

31. "Final Report: Pollutant Monitoring at the Picillo Dumpsite, Prepared for the Rhode Island Department of Environmental Management," S & D Engineering, Inc., East Brunswick, NJ, 1981.

32. Gioia, P. Unpublished results (1979).

6

COMPATIBILITY TESTING AND MATERIAL HANDLING

Mark A. Puskar and **Steven P. Levine**
School of Public Health
The University of Michigan
Ann Arbor, Michigan 48109-2029
Rodney D. Turpin and **Joseph P. Lafornara**
U.S. Environmental Protection Agency
Environmental Response Team
Woodbridge Avenue
Edison, New Jersey 08837

The stockpiling or dumping of waste chemicals in drums has been a widespread practice throughout the United States. At one-quarter of all abandoned waste sites major drum-related problems, including handling, integrity, characterization, and disposal exist [1,2]. Bulk recontainerization (bulking) of hazardous materials located in drums, labpacks, tanks, and holding ponds, has become the most time- and cost-effective solution to handling the waste.

Bulking involves removing the hazardous materials from the drums, labpacks, tanks, and holding ponds and combining compatible wastes into larger transportable containers [3]. When mixed together, many chemicals can produce potentially hazardous effects such as fires, explosions, violent reactions, and the release of toxic dusts, mists, fumes, and gases. Therefore, chemical wastes must first be tested for compatibility before they are bulked and incompatible chemicals must be segregated [4].

Chemical compatibility testing is not as extensive as the complete characterization of each individual drum of waste by standard laboratory testing procedures (e.g., RCRA Series, GC/MS, Emission Spectrography). Complete chemical characterization would be far too costly and time consuming for the purposes of hazardous waste cleanup. Compatibility testing, followed by bulking compatible drum contents into large containers appears to be the most cost- and time-effective method of preparing the waste for transportation to an approved disposal facility. Therefore, the second goal of compatibility testing is to generate enough knowledge of the characteristics of the waste to develop a hazardous waste classification for the shipping manifest which will be acceptable to the site manager, state or federal agency, transporter, and receiver of the waste.

This chapter will discuss the importance of compatibility testing, the general steps involved, and the limitations of compatibility testing. In general, compatibility testing involves performing a group of simple chemical tests (e.g., water solubility, pH, flammability) following a flowchart scheme which ultimately classifies the material into general categories [3,5-12].

IMPORTANCE OF COMPATIBILITY TESTING

The importance of compatibility testing is evident after examining situations where proper compatibility testing was not performed prior to handling of potentially hazardous waste.

On April 21, 1980, a massive explosion and fire occurred in Elizabeth, New Jersey, at a hazardous waste site owned by the Chemical Control Corporation. Of the 45,000 drums present at the site, more than 20,000 were consumed. During the initial site investigation which had been underway for 12 months prior to the fire, no compatibility testing and drum segregation had been performed. As of September 1981, the post-fire cleanup had cost more than $27 million, the costliest hazardous waste cleanup operation in history [13]. Final site mitigation has not yet been completed [14].

In January of 1982, workers were unloading chemical wastes from a tank truck at the Liquid Disposal Incineration, Inc., in Utica, Michigan. Deadly hydrogen sulfide gas was generated when a mislabeled sulfide waste from the tank truck mixed with acid in the receiving vessel. Two workers died and six others required hospitalization. Compatibility testing had not been performed on the incoming waste [15].

COMMON CHARACTERISTICS OF COMPATIBILITY METHODS

Although the above incidents exemplify the importance of compatibility testing, a universally accepted compatibility testing procedure does not exist. The contractor that is awarded the site cleanup contract is given a suggested scheme

to follow. However, they are encouraged to submit their own
drum consolidation protocols and analytical techniques. The
only requirements are that these proposed protocols provide
analytical data, compatibility and consolidation effective
and environmentally sensitive during removal and disposal of
hazardous waste from the site [8]. The scheme chosen for a
specific remedial action site is based on the following
factors:

(1) The general kinds of waste materials suspected to
be present on-site as determined by the initial
site screening.
(2) The criteria chosen by the governmental agency
supervising the cleanup.
(3) The preferences of the prime contractor and their
experience with compatibility testing.
(4) The criteria of the disposal facility to which the
waste will ultimately be sent.

The groups listed below have all developed compatibility
schemes. These schemes separate the waste into as few as ten
categories [9,12] and as many as 41 [3]:

1. USEPA, Office of Research and Development, Municipal
Environmental Research Laboratory (MERL) [3].
2. Samsel Services Company, Cleveland, Ohio [9].
3. Environmental Response Team, USEPA [10].
4. O. H. Materials Co., Findley, Ohio [12].
5. U.S. Army Corps of Engineers [8].
6. ASTM, scheme developed by committee D-34 [7].
7. Chemical Manufactures Association, Inc. [11].
8. NIOSH, Occupational Safety and Health Guidance
Manual [16].

The basic procedure in compatibility testing involves
subdividing the liquids into general disposal categories,
given the following assumptions:

1. A large number of drums exist on-site, and simple
overpacking with complete laboratory analysis
(GC-MS) is not time- or cost-effective.
2. An on-site facility is available with proper room,
equipment, and experienced personnel to perform the
analytical tests.
3. The waste contains a complex mixture of solids and
liquids.

Compatibility testing is only one step in the handling
of drum wastes on a remedial action site. Figure 1 diagrams
the role that compatibility testing plays during drum
handling.

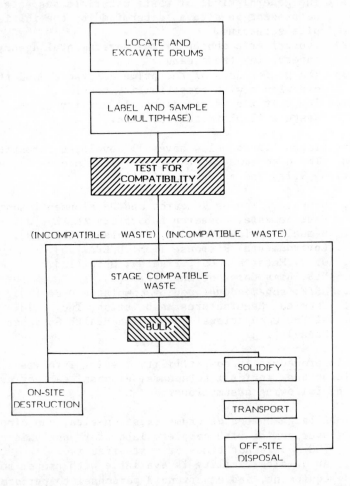

Figure 1. Flow diagram of drum handling operations during site remedial action.

Although each compatibility scheme is unique, most follow a similar flowchart. Figure 2 is the flowchart of the scheme used by O. H. Materials Co. Note that incompatible wastes such as radioactive, air and water reactive, PCB, sulfide, and cyanide wastes are identified by compatibility testing, but are usually not bulked prior to disposal. These highly hazardous wastes are repacked or overpacked, depending on the condition and size of their original container, and disposed separately.

The basic steps in compatibility testing occur in three stages of drum handling: 1) testing performed prior to drum opening; 2) testing performed during drum handling; and 3) testing performed on collected samples. The tests generally performed in each of these stages are discussed below.

TESTING PERFORMED PRIOR TO DRUM OPENING

Radioactive Wastes

The sampling team should identify radioactive wastes during the initial stage of site evaluation, using a gamma survey instrument. Following this procedure and performing the initial air monitoring will help ensure the safety of site personnel. Radioactivity levels are checked by scanning the closed drums. The OSC (On-Site Coordinator) should be notified immediately of all drums found having radiation levels above the local background. The OSC should then determine the appropriate handling and disposal steps for radioactive drums with the state or federal agency having responsibility for radioactive wastes.

All workers performing radioactive drum handling, sampling, or analysis should be monitored by documented radiation dosimetry techniques. Since normal environmental gamma radiation background is approximately 0.01 to 0.02 milliroentgen per hour (mR/hr.), routine employee exposure should not be more than 2-3 times this background level. At no time should routine employee exposures be at or above 10 mR/hr. without the advice of a qualified health physicist [10].

The absence of instrument readings above background should not be interpreted as the complete absence of radioactivity. Radioactive materials emitting low-level gamma, alpha, or beta radiation may be present, but for a number of reasons may not cause a response on the portable instrument. However, unless they are airborne, these

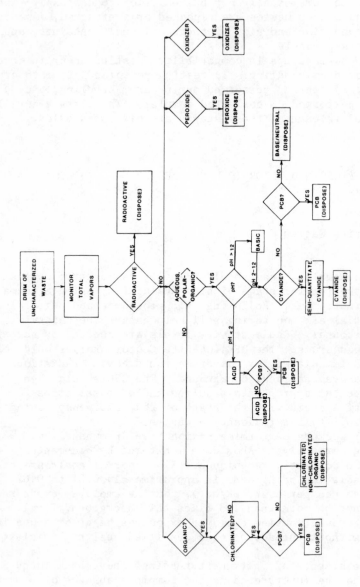

Figure 2. Compatibility testing for characterization of hazardous wastes.

radioactive materials should present minimal hazard. Reanalysis for radioactivity during sample compatibility testing is recommended.

TESTING PERFORMED DURING DRUM SAMPLING

Explosives/Air Reactives

After radioactive testing has been performed and all radioactive drums have been separated, the remaining drums are staged and opened for sampling. Drum opening, sampling, and staging protocols are shown in the process flow diagram in figure 1, and detailed in the literature [8,11] and must be strictly followed.

The objective of taking total vapor concentrations values just inside the bung hole is to assist in determining whether or not the headspace has a potentially explosive atmosphere. A potentially explosive atmosphere exists when a direct reading instrument reads at 2000 ppm (0.2%) or greater. An initial practice is to classify drums at or above 2000 ppm as flammable until further flammability data is available. The limitations and operating characteristics of the monitoring instrument must be recognized and understood. Instruments such as the photoionization detectors (PID) and the organic vapor analyzer (OVA-FID) have unique sensitivities and specificities to identical substances, and proper calibration when dealing with "unknowns" is impossible. Also, dangerous gases/vapors undetectable by photo and flame ionization detectors may be present. Such gases would include: phosgene, hydrogen cyanide, chlorine gas, and liquid/solid particulates.

The next compatibility test performed on opened drums is air reactivity. This is performed by visual observation during sampling. Any sample taken from a drum containing a solid which ignites or emits fumes, gases, or releases pressure is considered air reactive and immediately separated [9]. Also, any drum found containing metal submerged in liquid should be immediately segregated and considered air reactive. Additional sampling should be done to identify the type of metal. The most common elements found disposed in this manner are phosphorus (air reactive) and sodium (water reactive).

TESTING PERFORMED ON COLLECTED SAMPLES

Water Reactivity/Solubility

Radioactive, air reactive, and explosive testing are performed prior to and during sampling. One of the first compatibility tests performed on the collected samples should be water reactivity/solubility. This simple test can generate a host of information on the uncharacterized waste.

The characteristics of reactivity, as defined in the RCRA regulations (40 CFR 261.23), are exhibited if a representative sample of the waste has any of several properties, including:

(1) it reacts violently with water;
(2) it forms potentially explosive mixtures with water;
(3) when mixed with water, it generates toxic gases, vapors, or fumes in a quantity sufficient to present a danger to human health or the environment.

The compatibility methods call for a small volume (1 mL for highly reactives to 10-mL for non-reactives) of liquid waste to be added to water and the mixture observed for water miscibility, temperature exotherm, precipitation, and/or gas formation. If any of these occur, the waste is classified water reactive. Following this definition, acids and bases are initially classified as water reactive. They will later be separated by pH measurements. A major interference of this test is that certain water reactive materials may require a reaction time, catalyst, or heat before reactions occur. While potentially dangerous, this test can be relatively safe if precautions against explosions and toxic vapors hazards are taken. Unknown organic vapors, hydrogen cyanide, hydrogen sulfide, chlorine, ammonia, and hydrogen gases could be generated in small amounts.

Another scheme [9], only classifies materials which evolve gases, fumes, or ignite as water reactive. Materials which produce a temperature change are not considered to be water reactive unless the water approached boiling temperature. Following this definition, acids and bases are initially separated from water reactives. Water reactive drums should be isolated and sheltered from the elements. However, because of the danger of explosion and fire, indoor storage is not recommended.

Organics/Inorganics

When a liquid sample from an uncharacterized waste drum is placed in water and is nonreactive, it will either be soluble or insoluble in the water. If it is insoluble, it will either sink or float (becomes the top or bottom phase).

Samples that are soluble in water are strong suspects for inorganic classification. The solubility of these samples is determined in hexane, and if they are soluble in the hexane they are classified as nonhalogenated organics (polar). If they are insoluble in the hexane, they are classified as inorganic liquids.

Organics/Halogenated Organics

Samples that are insoluble in water and are the top phase are classified as organic liquids. Two schemes [3,11], check the vapor concentration above the sample at this point with either a PID or a OVA-FID to determine if the organic liquid should be classified as volatile or not. A value of |2000 ppm is recommended to classify an organic sample as volatile.

Samples that are insoluble in water and are the bottom phase are classified as halogenated organics. This procedure for determining halogenated organics is essentially a procedure designed to determine gross halogen content, not parts per million as required by law for PCBs. Further tests to confirm the presence of halogens using either a copper loop flame test [12], halogenated organic GC-ECD scan [18], potentiometric titration [11], and total organic halogen (TOX) have been suggested.

Analysis of all drums classified as organic for PCBs [8] must be conducted to determine if ultimate disposal is possible in a non-PCB approved incinerator (for liquids) or landfill (solids and soil). This testing procedure, semi-quantitative from a hexane extraction by gas chromotography [11], should be conducted prior to drum bulking so isolation of PCB-contaminated containers and/or a change in bulking sequence can be inserted to avoid further PCB contamination. The PCB analysis can be conducted on composited samples, to save time and money; however, it would be expedient to not composite more than ten drums per sample. Any drums determined to be > 50 ppm PCB are classified as PCB-contaminated wastes.

Flammability

Although a standard flashpoint test would furnish helpful information about the unidentified waste, this analytical method becomes impractical when dealing with large numbers of drums on a remedial action site. For example, if a test could be performed every 20 minutes, it would take approximately three man-years to analyze 20,000 mono-phasic drums. Thus, simple flammability techniques have been developed to separate flammables from nonflammables.

For solid unknowns, a small sample (20 to 50 mg) is transferred on a steel spatula into an open flame. If the material is a liquid, a stainless steel loop holding a drop of the unknown is employed [9]. If the sample ignites violently at some point in the heating, it is classified as an explosive. If the material burns with the flame, it is considered flammable and organic. Exceptions to this guideline are inorganics such as phosphorus and sulfur; however, their characteristics during burning may be used to distinguish them from organics. Also, clear halogenated organics, such as chloroform or carbon tetrachloride, test as inorganics in the flame test. If the material does not burn, but chars or turns black, it is suspected of being organic. If the material does not burn, it is considered inorganic.

Another scheme [10] suggests placing a 2-5 ml representative sample in a disposable beaker. The beaker is placed in a large sandbox and a propane torch is slowly passed over the unidentified waste. If a flame is observed, the waste is classified as flammable. A nonflammable classification is assigned to the waste after the torch has passed over the waste several times. Certain disposal facilities may require confirmation flashpoint testing to be conducted on all spot-tested samples which were classified positive.

Most observers will be able to make a distinction between organics, inorganics, and free metals using a flame test. Because of the dangers of explosions or fires with these types of flammability testing, safety requirements must be strictly followed. Indoor flammability testing should be performed in explosion-proof high-flow hoods. Fire extinguishers and safety showers must be immediately available.

pH

To guard against explosive exothermic reactions and evolution of deadly gases (cyanide, sulfide) caused by the commingling of caustic and acidic wastes, pH measurements are taken to separate potentially dangerous drums. The pH of each sample, previously classified as water soluble, is determined using either an electronic pH meter with temperature compensation adjustment and appropriate electrodes, or indicator strips (pH paper) covering the pH range of interest. Both have disadvantages when used on "dirty" samples containing organic layers, sludge, or concentrated solutions. For instance, standard pH electrodes are easily fouled and require constant cleaning, recalibration, and regular replacement. Most colorimetric indicators and papers are easily obscured by grease, sludge, or opaque solutions. Interfering chemicals may even cause false color changes. These interferences are limited by the use of multiband pH paper which contains reaction zones and a series of indicator colors fixed for reference. After the strip has been exposed to the waste, the color comparison of the reaction zone and the indicator color is made "assuming" that both have been affected in the same way by the waste.

The greatest concern with the commingling of acids and bases is the generation of deadly cyanide and sulfide gases. Cyanide and sulfide wastes are usually buffered at a pH of 10 in order to remain in aqueous solution. This important fact has been used to define acids and bases for compatibility purposes. Caustic wastes are defined as those with a pH above 10, and acidic wastes are defined as those with a pH below 10 [10]. Following this definition, the accidental release of sulfide or cyanide gas during drum bulking is greatly reduced.

Other schemes [8,9,11] separate the wastes into three or four categories depending on the pH value. Wastes with pH values less than 2 are classified as acids; between 2 and 7 are classified as acidic aqueous solutions; between 7 and 12 are classified as basic aqueous wastes; and greater than 12 are classified as bases.

Wet methods [11], and the use of ion-selective electrodes [12], are used to determine the presence and concentration of cyanide and sulfide in bases. All drums tested positive are separated from the bases. Additional staging safeguards, including barriers, are recommended to reduce the risk of accidental mixing of cyanide/sulfide bases with acids.

The method of defining acids and bases and selecting analytical methods should be based on the future analytical tests required to meet the criteria set by the approved disposal facility accepting the waste. Many disposal facilities require base and acid reactivity testing prior to acceptance. This method can be found in the Federal Register, 40 CFR 261, Subpart C.

Strong acids and (sulfide- and cyanide-free) bases identified can be blended on-site for neutralization. These reactions will be highly exothermic and should not be attempted without adequate safeguards [11] including real-time sulfide/cyanide air monitoring systems and bulking chamber temperature monitors. On-site neutralization may significantly lower disposal costs of waste acids and bases.

Oxidizers/Reducers

Compatibility testing procedures have been developed for analyzing and classifying drums containing oxidizing or reducing agents, including wet methods, test papers, and portable instrumentation methods.

The wet methods [9] include a colorimetric determination of organic peroxides in solid organic unknowns using titanium sulfate as a yellow color indicator and a colorimetric determination of inorganic oxidizers in solids or liquids using manganous chloride as a black or brown indicator. Test papers have had the widest acceptance due to their quickness and simplicity [3,12]; however, the authors are unaware of any documentation of their ability to determine oxidizers and reducers in complex waste samples.

A potentiometric determination of the redox potential of drum samples through the use of a portable battery-operated instrument has been developed and tested [10]. The unique features of this method is its ability to perform redox measurements not only in aqueous but also organic matrices, such as are found on hazardous waste sites. The entire procedure requires only a few minutes and can be performed by inexperienced operators in the drum staging area. The test is very sensitive and a reaction with only a small portion of an oxidizing agent will give a positive test.

The method involves using an electrolyte solution to generate a known redox potential and then measuring the change in the potential when an unknown waste is added to the electrolyte. Ferrous ammonium sulfate, as a standard electrolyte, is used for oxidation readings. Potassium chromate is substituted for reduction measurements.

Two minor problems have been encountered with this field procedure: electrode probe clogging and electrolyte freezing at sub-0 C temperatures. The clogging problem is identical to the pH electrode clogging problem and can be resolved by proper cleaning of the probe between samples. The cold weather, however, causes the probe electrolyte to freeze. This problem is resolved by conducting the tests inside an on-site laboratory.

Labpacks are 55-gallon drums which contain small volume containers of waste chemicals. Most are unlabeled chemical reagents discarded by laboratories. Proper disposal of this type of container presents a unique and hazardous clean-up problem. The chemicals stored in many of these small containers are incompatible. Usually, these containers were not packed in absorbent material prior to original disposal, so breakage and chemical mixing during drum handling is common. For this reason, these drums are considered primary ignition sources for fires during remedial action.

Compatibility testing is performed by chemists conducting visual inspection of the containers, without opening, and attempting to classify them. If crystalline material is observed in the neck of any bottle, it is handled as shock sensitive, due to the potential presence of picric acid or similar materials. Shock-sensitive containers are repacked in absorbent material, and repacked not more than five to a drum, and shipped to a disposal facility or detonated on-site. The containers with markings that can be identified and trusted (a purely subjective judgement) are segregated into similar compatibility categories and repacked in absorbent material [17].

The repacking protocol for unidentifiable containers is set by the facility accepting them for final disposal. This usually involves separating the unknowns into solids and liquids, and the liquids into single- and multi-phased.

A compatibility scheme for labpacks has been tested which identifies expected chemicals among unlabeled materials, segregates the remaining unlabeled materials into one of six disposal groups and destroys or neutralizes the excepted materials sufficiently to permit landfill disposal. This method [9] involves opening, sampling, and performing classic compatibility testing on each container.

Another suggested scheme for the disposal of labpacks calls for opening each container with a high velocity, low mass projectile (a 0.22 caliber bullet) [19]. This safely accomplishes two things: the container (almost exclusively glass bottles) are remotely opened by fracture from the bullet impact, and the contents are collected and rendered stable in an absorbent material. Advantages to this method include simplicity over remote control detonation, minimal set-up, low-technical procedure, low visibility compared to detonation, and low cost; however, this procedure would be impossible in a populated location. In highly populated locations, this scheme has been adapted by opening each container by running it over with a bulldozer, and then performing RCRA testing on the contaminated soil

Limitations of Compatibility Testing Procedures

It should be understood that the compatibility methods discussed are designed to detect acute incompatibility only. Reactions which require heat or other catalyzing effects for initiation will not be detected by these tests [8].

Mixtures are only identified as far as it is necessary to place them into one of the disposal groups. Since these procedures are not designed to identify specific compounds, there are a number of types of materials for which these procedures are not applicable. The identification of highly toxic or carcinogenic materials, such as dioxin, which should not be landfilled is beyond the scope of these procedures. Although these procedures may be employed to segregate total unknowns into disposal groups, additional specific tests must be performed to determine the presence and concentration of materials that cannot be landfilled [9]. This problem may be controlled for by incinerating all nonsoil wastes in a Class B facility; however, this solution is not cost-effective.

The need to test-mix the wastes on a small scale, prior to drum bulking, is emphasized even if the compatibility tests indicate compatibility. A major problem with the test mixing, and thus compatibility testing, is obtaining a homogeneous sample from each uncharacterized drum, especially when working with multi-phased drums containing both liquids and solids. Even after the unknown drums are mixed, the clean-up personnel are unable to estimate the short-term (minutes to hours) or long-term (days to weeks) effects of mixing and recontainerizing unknown materials [3].

Very little data exists that document the effectiveness of compatibility schemes in separating unknown wastes. The data that does exist details major classes that cannot be identified, including isocyanates, epoxides, nitriles, and polymerizable materials. In addition, a need exists for a sound quality-control criteria for compatibility testing [3].

No methods have been proposed for the screening of pathogenic or infectious materials. These may be present in the uncharacterized drums, especially labpacks.

Although compatibility testing is generally detailed enough for disposal purposes, its use from an industrial/environmental hygiene standpoint is limited. Since no other assays are performed (unless expensive GC-MS is ordered for legal purposes), very little information exists, particularly on drum content, that is useful from a toxicological standpoint. For example, when a drum is categorized non-halogenated organic, no other data is generated to determine if the drum contains a slightly toxic compound toluene (TLV 100 ppm, 8.7 g/kg LD50), a moderately toxic chemical benzaldehyde (TLV None, 1g/kg LD50), a highly toxic chemical parathion (TLV 0.01mg/m3, 2mg/kg LD50), or a mixture of all three.

A greater knowledge of drum material composition would permit the tailoring of worker and community protection strategies to those specific materials. To meet these needs, however, classic compatibility test procedures (spot tests) have become more and more complex. Consequently, they have lost their time- and cost-effectiveness and become prone to positive and negative interferences [3].

A need exists for an analysis procedure that can potentially fill the gap between compatibility testing and expensive GC-MS. This procedure should:

(1) Furnish chemical information beyond that which is obtainable via compatibility testing, (i.e., identification of the primary constituents of uncharacterized drum samples);

(2) be rapid enough to complete 100-200 samples with a 24-hour turnaround;

(3) be cost-effective with respect to compatibility ($30/sample) and simple GC-MS ($750/sample) testing.

Research continues in the development of new compatibility testing procedures, which attempt to meet the criteria listed above; however, until a new method is developed and tested following proper peer review, the classical spot-test methods are recommended as the only alternatives.

SUMMARY

This chapter addressed the role that compatibility testing plays during the waste characterization and bulking phase of a remedial action. The classical spot-test methods were discussed in depth, and proposed unique methods were outlined. Major safety problems encountered and proposed safety protocols were detailed.

The limitations of compatibility testing was addressed, pointing out the need for a quality-control protocol. A criteria based on the need for fast, cheap, but high quality compatibility testing was given against which to judge future proposed methods.

REFERENCES

1. Wetzel, R., et al. "Drum Handling Practices at Abandoned Sites," Management of Uncontrolled Hazardous Waste Sites Conference, Washington, D.C., 1982. Bennett and Bernard, eds., Silver Spring, Maryland, Hazardous Waste Control Research Institute, 1982.

2. Neely, N., et al. "Remedial Action at Hazardous Waste Sites. Survey and Case Studies." EPA 430-9-81-005. SCS Engineers, Covington, KY for USEPA Municipal Environmental Research Laboratory, Cincinnati, OH, 1981.

3. Wolbach, C. D. "Protocol for Identification of Reactivities of Unknown Wastes," Management of Uncontrolled Hazardous Waste Sites Conference, Washington, D.C., December 1983. Bennett and Bernard, Eds., Silver Spring, MD, Hazardous Waste Control Research Institute, 1983.

4. Federal Register, 40 CFR, Part 260, May 19, 1980.

5. "Hazardous Material Incident Response Operations Training Manual," U.S. EPA Office of Emergency and Remedial Response, Hazardous Response Support Division, June 1982.

6. "Hazardous Material Response Manual," U.S. Coast Guard, Environmental Coordination Branch, Draft Document, May 1982.

7. "Guide for Determining the Compatibility of Hazardous Wastes," ASTM D-34.04.04, Draft 3, June 10, 1982.

8. "U.S. Corps of Engineers/U.S. EPA Hazardous Waste Site Remedial Action Guidelines," RFP for Chem-Dyne Site, Hamilton, OH, January 1983.

9. Hina, C. E., et al. "Techniques for Identification and Neutralization of Unknown Hazardous Materials," Management of Uncontrolled Hazardous Waste Sites Conference, Washington, D.C., 1983, Bennett and Bernard, Eds., Silver Spring, MD, Hazardous Waste Control Research Institute, 1983.

10. Turpin, R. D., et al. "Compatibility Testing Procedures for Unidentified Hazardous Wastes," Management of Uncontrolled Hazardous Waste Sites Conference, Washington, D.C., 1981, Bennett and Bernard, eds., Silver Spring, MD, Hazardous Waste Control Research Institute, 1981.

11. Mayhew, J. D., G. M. Sodaro, and D. W. Carroll. A Hazardous Waste Site Management Plan, (Washington, D.C.: Chemical Manufactures Association, 1982.

12. "Compatibility Testing for Characterization of Hazardous Waste," O. H. Materials Co. (1983).
13. Finkel, A. M., and R. S. Golob. "Implications of the Chemical Control Corp. Incident," Management of Uncontrolled Hazardous Waste Sites Conference, Washington, D.C., 1981, Bennett and Bernard, eds., Silver Spring, MD, Hazardous Waste Control Research Institute, 1981.
14. "Hazardous Waste Sites: National Priorities List," U.S. EPA, U.S. Government Printing Office (1983).
15. Clinical Outpatient Notes, Occupational Health Clinic, The University of Michigan, September 22, 1982.
16. "Occupational Safety and Health Guidance Manual," NIOSH, Appendices A-C1, Draft Copy (1983).
17. Wyeth, R. K. "The use of Laboratory Screening Procedures in the Chemical Evaluation of Uncontrolled Hazardous Waste Sites," Management of Uncontrolled Hazardous Waste Sites Conference, Washington, D.C., December 1981, Bennett and Bernard, eds., Silver Spring, MD, Hazardous Waste Control Research Institute, 1981.
18. "Lab Pack Disposal Procedures," U.S. EPA (1983).
19. "Request for Modification to Approved Work Plan for Handling Lab Packs," O. H. Materials Co. (1983).

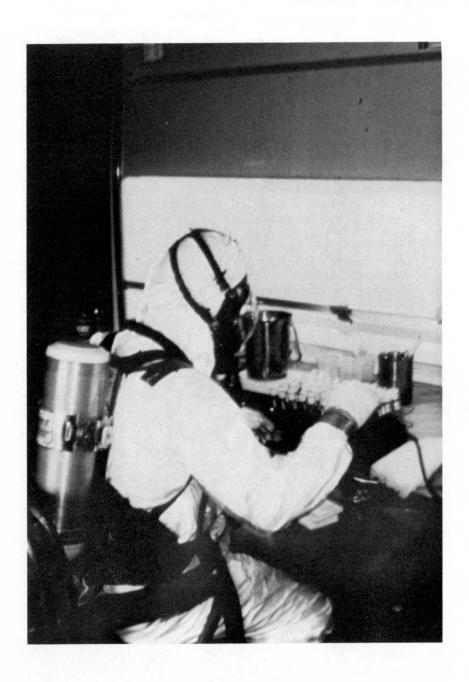

MEDICAL SURVEILLANCE FOR HAZARDOUS WASTE WORKERS

James M. Melius, M.D.
Chief, Hazardous Evaluation and Technical Assistance Branch
Division of Surveillance, Hazard Evaluations and Field Studies
National Institute for Occupational Safety and Health
Cincinnati, Ohio 45226

Designing a medical surveillance program for workers is a difficult task [1]. At the present time, there are very few quantitative data about the potential occupational exposures and hazards facing these workers. Available industrial hygiene data indicate that these workers have low-level exposures to multiple chemicals and the possibility of short-term high-level exposures to many of these chemicals. The latter exposure situations provide the rationale for much of the protective equipment and work practice programs described elsewhere in this book.

Most occupational medical screening is based on the known toxic effects of specific chemicals and is predicated on the occurrence of a significant degree of exposure to those toxic substances. Even for specific toxic substances, most medical screening recommendations have not been critically evaluated for efficacy. Combining the usual medical screening recommendations for each chemical to which the hazardous waste worker could be exposed would produce a costly, unwieldy list of screening recommendations that would be of doubtful effectiveness.

The following recommendations for a medical program for workers are based on the established health hazard for those workers, a review of the available data on their exposures, and a review of several established occupational medical programs for hazardous waste workers. Currently, a variety of approaches have been taken. This chapter is intended not to criticize individual programs, but rather to show the basic steps in designing medical programs for different hazardous waste cleanup operations (see Appendix). The medical recommendations are intended for a program under the direction of a physician trained in occupational health or with considerable experience in conducting occupational health programs. These recommendations also are based on the assumption that workers will be adequately protected by the use of engineering controls and work practices as delineated elsewhere in this book. The recommendations are presented in four parts: preemployment screening, periodic screening, provisions for episodic and emergency medical care, and recordkeeping.

PREEMPLOYMENT SCREENING

The major focus of preemployment examinations should be to ascertain whether the worker is physically fit to perform the assigned work. This work often involves physically strenuous activity (moving 55-gallon drums, etc.) and, in addition, requires the worker to wear personal protective equipment (respirators, protective suits, etc.). Wearing this equipment poses an added physiological burden on the worker, particularly when working in high ambient temperatures. Unfortunately, there is no accurate method of quantitatively measuring this added physiological burden at the present time [2]. The preemployment screening should therefore include a medical history and physical examination to determine if the worker will be able to handle strenuous work while wearing personal protective equipment.

The Occupational Safety and Health Administration (OSHA) respirator standard (29 CFR 1910, Part 134), states that no employee may be assigned to a task that requires the use of a respirator unless it has been determined that the person is physically able to perform under such conditions. The medical history should ascertain information on past illnesses and chronic diseases (particularly asthma, pulmonary disease, and cardiovascular disease), and include a review of symptoms (especially dyspnea on exertion, other chronic respiratory symptoms, chest pain, and heat intolerance). Other characteristics which may make an individual more susceptible to heat stroke, such as obesity and little physical exercise, should also be ascertained. The physical examination should focus on the pulmonary and cardiovascular system. Depending on the results of the medical history and physical examination, and on the worker's age, further medical testing, such as a chest X-ray, pulmonary function testing, and an electrocardiogram may be useful in ascertaining the person's ability to perform

strenuous work while wearing a respirator and other
protective equipment. These additional tests, however, need
not be done for everyone. The medical history and exam by
themselves may disqualify some individuals. Little
information would be gained from these additional tests for a
young, healthy, nonsmoking worker. On the other hand,
pulmonary function testing and an electrocardiogram may prove
quite useful in evaluating an older worker with a long
history of cigarette smoking.

Based on the medical history, physical examination, and
appropriate further tests, the examining physician must then
make a decision on the worker's ability to perform the
required work while wearing protective equipment.
Unfortunately, there is very little sound guidance in the
literature on which to base this decision [2-8]. Prospective
employees with severe lung or heart disease should obviously
be excluded; however, there are no clear-cut guidelines for
an asymptomatic worker with modest reduction in pulmonary
function. Usually in these instances the medical assessment
must be based on an overall assessment of the person's
medical examination. Current research on the physiological
burden involved in wearing respirators and protective
clothing, and on the effects of reduction in pulmonary
function and respirator tolerance, should help provide better
guidance for these assessments in the future.

Another major purpose of preemployment testing is to
ascertain the worker's ability to work in hazardous
environments (i.e., is the worker unusually susceptible to
specific chemical exposures?). Since exposures at hazardous
waste sites are multiple and often unpredictable, other than
serious medical conditions which would disqualify a worker by
the above criteria or a history of severe asthmatic reaction
to a specific chemical, specific medical testing would not be
effective for this purpose.

The final purpose of preemployment medical screening is
to establish baseline data to better evaluate the effects of
subsequent toxic exposures. This baseline testing may
include both medical screening tests and biological
monitoring tests. The latter (e.g., blood lead level) may be
useful for ascertaining preexposure body burdens of specific
substances to which the worker may be exposed and for which
reliable tests are available. Given the problem in
predicting significant exposures for these workers, there are
no clear guidelines for prescribing certain tests.
Alternative approaches range from doing no testing to
conducting an extensive battery of biochemical and biological
monitoring. A more rational approach would include baseline
testing selected according to the past history of the workers
(previous medical and occupational history) and on some
assessment of the predominant and significant exposures which
the worker may experience.

149

The most common potential chemical exposures for workers at a hazardous waste site are solvents. Although some solvents have specific toxicity (e.g., leukemia can result from benzene exposure), the most common medical effects from solvent exposure are neurotoxic and hepatotoxic. Other than history and physical examination, routine preemployment screening tests for neurotoxic effects are not available. Liver enzyme tests are commonly used in testing for hepatotoxic effects, but their sensitivity and specificity for detecting the effects from low-level exposures to multiple solvents are unknown and probably low [9]. Their use in preemployment or periodic screening is questionable. Likewise, the value of other commonly used preemployment biochemical tests (BUN, calcium, etc.) or other baseline laboratory testing is also questionable.

An alternative approach to baseline testing would involve a situation where a specific significant exposure for the hazardous waste worker is known and biological or biochemical monitoring of that exposure is well established. For example, long-term cleanup of a polychlorinated biphenyl (PCB) waste facility could be monitored with preemployment and periodic serum PCB testing [10]. Lead, cadmium, arsenic, and organophosphate pesticides are additional examples of substances for which this approach could be appropriate. Given the common use of respirators and protective clothing for workers at hazardous waste sites, usual industrial hygiene monitoring will not provide an accurate indication of the worker's actual exposure (i.e., through the respirator and protective clothing). Therefore, in situations where a hazardous waste worker may be exposed over a sufficient period of time to a substance which can be monitored by available biological monitoring techniques, preexposure and periodic biological monitoring for that substance may provide very useful information on the actual exposure for that worker or group of workers.

A related approach would involve drawing preemployment blood specimens and freezing serum for later testing if environmental monitoring indicates significant exposures to an agent amenable to such monitoring (e.g., PCB, some pesticides).

PERIODIC SCREENING

The frequency and content of periodic screening of hazardous waste workers will depend on the nature of their work. In general, these examinations should take place at least yearly and include an interval history and physical

examination. The medical history should focus on changes in
health status, illnesses, and possible work-related symptoms
occurring since the last screening examination. The
examining physician should have some knowledge of the
worker's exposure during that period of time. This
information should include any exposure monitoring done at
the worker's job site. This could be supplemented by
self-reported exposure histories or more general information
on the potential exposures at the hazardous waste sites where
the employee has worked. Additional medical testing would
depend on the available exposure information and on the
medical history and examination results. This additional
testing should be specific for the possible medical effects
of the worker's exposure. The application of a large batch
of medical tests in an attempt to cover all of the possible
medical effects of the multitude of potential exposures
facing the worker is not very useful. Such testing may only
lead to problems due to the occurrence of elevated values due
to other factors or to chance (i.e., false positives).

More frequent monitoring may be appropriate for
significant exposures at specific sites (e.g., PCB, lead,
etc.) as described in the preceding section. The schedule
for this monitoring would depend on the degree and type of
exposure and the duration of work at the job site. Periodic
review of the screening results can help determine the
appropriate frequency. For example, workers involved in the
cleanup of a building contaminated by PCB initially had
monthly serum PCB levels [9]. Review of data from the first
few months revealed no evidence of appreciable PCB exposure,
thus, the frequency of PCB testing was reduced accordingly.

ACUTE OR EPISODIC MEDICAL CARE

Provisions for acute medical care need to be developed
for each hazardous waste site. This should include
provisions for emergency first aid at the site. Key
employees at the site should have some formal first aid
training, particularly in dealing with explosion and burn
injuries, with heat stress, and with acute chemical
toxicity. Appropriate first aid equipment also needs to be
available at the site.

Arrangements for evacuating injured or ill personnel
need to be available, including transportation to a nearby
hospital. These arrangements should include assisting the
hospital in preparing for medical occurrences. Preparation
can avoid unnecessary delays in treating injured or ill
workers due to inappropriate concerns about chemical

contamination of the hospital (this has actually occurred after a fire at a hazardous waste facility). The medical care facility should be informed about the nature of potential exposures at the site, the specific details on the incident involving the ill or injured worker, and about the worker's medical history. These arrangements are particularly important when specific medical treatment is required for a toxic exposure (e.g., cyanide, organophosphate pesticides).

In addition to the provisions for a medical emergency, a mechanism to provide episodic medical care for hazardous waste workers needs to be arranged. This may be difficult, particularly if the worker is not close to the home office of the employer or is working in a rural area. Nevertheless, it is important to ensure that any possible symptoms or illnesses are properly evaluated in the context of the worker's exposures at the site and that other illnesses do not put the worker at greater risk due to the requirements of working with hazardous waste. Arrangements need to be made for the treating physician to have access to the worker's medical records. Depending on the situation, this can be done by keeping the medical records (or a copy) at the hazardous waste site (with appropriate provisions for security) or at a nearby hospital.

Another important group of workers who may be exposed at hazardous waste sites is emergency response personnel. These workers may encounter significant acute exposures in responding to fires or other emergencies. Proper preparation can help prevent serious exposures in these situations. Prior to the hazardous waste site cleanup, the fire department and other emergency response personnel need to be informed on potential hazards from incidents at the site. Procedures to limit these exposures and to assure the availability of appropriate protective equipment can be made. Arrangements also need to be made for decontamination or disposal and replacement of fire fighting equipment used at the site. In the event of significant exposure for any of these workers, arrangements need to be made for appropriate medical or emergency care, including informing the medical care provider of possible exposures at the site.

RECORDKEEPING AND PROGRAM REVIEW

Recordkeeping is an important part of any medical surveillance program. For hazardous waste workers, this may be difficult due to the multiple locations where they may work over a period of time. Current OSHA regulations require that medical records on exposed workers be maintained for 30 years after they leave employment (45 FR 35212). The results of medical testing and full medical records must also be available to the workers, their union representatives, and OSHA inspection staff. Informing workers about their exposures and medical testing is particularly important in helping them to take appropriate precautions and for informing their other or subsequent medical care providers of their exposures as a hazardous waste worker. Occupational accident and illness records must also be maintained and reported yearly to OSHA.

A successful health and safety program should include periodic reassessment of its effectiveness. This will involve reviewing medical records; therefore, these documents should be well maintained. Each accident or illness should be promptly evaluated to determine the cause of the incident and to implement appropriate changes in the health and safety procedures for the site. This activity is particularly important in conducting a program for hazardous waste sites where the nature of the work and the variety of potential occupational exposures require good compliance with work procedures (e.g., respirator use) to maintain an effective health and safety program.

Periodic review of the results of medical surveillance testing is also important in maintaining the effectiveness of the medical surveillance program. This review should attempt to critically evaluate the efficiency of specific medical surveillance testing, particularly in the context of information on the exposures or potential exposures at hazardous waste sites. Industrial hygiene and environmental data may suggest the need for adding specific medical tests to the surveillance program. The director of the medical surveillance program should also review the potential exposures at new hazardous waste sites to determine if additional medical testing is required for the workers at that specific site.

CONCLUSION

The design and conduct of a medical program for hazardous waste workers is a difficult task. These workers are potentially exposed to thousands of toxic substances, often in situations where identification or quantification of these exposures is not possible. The medical program for these workers must provide a baseline of preemployment and periodic screening, yet remain adaptable to the exposures at specific sites. Most important, this program must be integrated with the industrial hygiene program, personal

153

protective equipment program, and safety procedures for the
site. Together these programs can provide a safe and
healthful workplace in what initially appears to be an unsafe
workplace--a hazardous waste site.

PREEMPLOYMENT SCREENING

A. Recommended

1. Medical history and physical examination with selective medical testing (e.g., chest X-rays, pulmonary function testing, EKG) to determine worker's fitness to work while wearing protective equipment.
2. Preemployment or (preexposure) baseline biological monitoring for specific exposure at a hazardous waste site (e.g., PCB).

B. Optional

1. Freezing a preemployment serum specimen for later testing.
2. Other routine baseline tests—blood count, liver enzyme tests, etc.

PERIODIC SCREENING

A. Recommended

1. Yearly medical history and physical examination with appropriate medical testing selected on the basis of this examination and on the worker's exposure history.
2. More frequent screening based on exposure to specific hazards (e.g., organophosphate pesticides, PCBs) or individual health factors.

B. Optional

1. Yearly testing using routine medical tests (e.g., blood count, liver function tests, etc.).

ACUTE MEDICAL CARE

A. Recommended

1. Provisions for emergency first aid at the site.
2. Provisions for hospital transportation, and for informing the hospital about exposures at the site, particularly if specific medical treatment is available for a toxic exposure (e.g., cyanides).
3. Mechanism for episodic health care with evaluation of possible site-related illness.

RECORDKEEPING AND PROGRAM REVIEW

A. Recommended

1. Maintenance and access to medical records in accordance with OSHA regulations.
2. Recording and reporting of occupational injuries and illnesses.
3. Periodic review of the medical surveillance program including integration with available exposure information about the hazardous waste sites where the workers are employed.
4. Review of specific site safety plans to determine if special testing is required for the workers at that site.

REFERENCES

1. Melius, J. M., and W. E. Halperin. "Medical Screening of Workers at Hazardous Waste Disposal Sites," in Hazardous Waste Disposal: Assessing the Problem, J. Highland, Ed. (Ann Arbor: Ann Arbor Science Publishers, 1982).

2. Raven, P. B., A. Jackson and K. Page. "The Physiological Responses of Mild Pulmonary Impairment While Using a 'Demand' Respirator During Rest and Work," Am. Ind. Hyg. Assoc. J. 42:247-257 (1981).

3. Petsonk, L., C. Boyles and T. Hodous. "Effects of Added Resistance to Breathing in Obstructive Lung Disease--Phase I," U.S. DHEW, NIOSH Report, Clinical Investigations Branch (1981).

4. Hodous, T., L. Petsonk and C. Boyles. "Effects of Added Resistance to Breathing in Obstructive Lung Disease--Phase II," U.S. DHEW, NIOSH Report, Clinical Investigations Branch (1982).

5. Hodous, T., C. Boyles and J. Hankinson. "Effects of Added Resistance and Dead Space on Breathing in Normals and Subjects with Restrictive Lung Disease--Phase II," U.S. DHEW, NIOSH Report, Clinical Investigations Branch (1982).

6. Pritchard, J. A. "A Guide to Industrial Respiratory Protection," U.S. DHEW, Publication NIOSH 76-189 (1976).

7. Raven, P. B., A. T. Dodson and T.O. Davis. "The Physiological Consequence of Wearing Industrial Respirators: A Review," Am. Ind. Hyg. Assoc. J. 40:517-534 (1979).

8. Tuomi, T. "Physiological Effects Associated with the Wear of Respirators," paper presented at the American Industrial Hygiene Association Conference, Philadelphia, June 1983.

9. Liss, G., C. Tomburo and A. Greenberg. "The Use of Serum Bile Acids in Medical Screening for Hepatotoxicity," paper presented at the American Public Health Association Conference, Montreal, November 1982.

10. Chase, K. "Medical Surveillance of Clean-up Workers at the Binghamton State Office Building," paper presented at the Expert Panet Meeting, Binghamton, NY 1982.

SITE LAYOUT AND ENGINEERED CONTROLS

Lynn P. Wallace, Ph.D., PE
Associate Professor of Civil Engineering
Brigham Young University
Provo, Utah 84602

The manner by which a hazardous waste site is managed will, to a large extent, affect the health and safety of not only personnel who work at the site, but also surrounding environments. Remedial actions are initiated to establish proper management at a site where evidence indicates it is needed.

Since on-site activities at hazardous waste sites usually include the investigation, handling, and movement of hazardous materials; personnel, equipment, and previously uncontaminated areas may become contaminated by spilled toxic materials or harmful airborne dusts and vapors if proper control is not established and maintained. Contaminants must be made harmless or contained and hazards must be eliminated or isolated to protect workers and others from injury, illness, or death, and to protect surrounding environments.

This chapter discusses site management as it relates to: 1) site layout, and 2) engineered controls.

SITE LAYOUT

The manner in which the site is physically arranged and laid out to accomplish the remedial objectives is one of the most important aspects of site management.

A properly laid out and managed site, such as one having excellent materials handling operations and control over entry and exit of all personnel, equipment, and materials, can significantly increase the health and safety of personnel on-site and contribute to successful remedial

actions. An EPA Training Manual offers the following [1]:
  "A site must be controlled to reduce the possibility
of: 1) exposure to any contaminants present and 2) their
transport by personnel or equipment from the site. The
possibility of exposure or translocation of substances can
be reduced or eliminated in a number of ways, including:

* Setting up security and physical barriers to exclude
  unnecessary personnel from the general area.
* Minimizing the number of personnel and equipment
  on-site consistent with effective operations.
* Establishing work zones within the site.
* Establishing control points to regulate access to
  work zones.
* Conducting operations in a manner to reduce the
  exposure of personnel and equipment and to eliminate
  the potential for airborne dispersion
* Implementing appropriate decontamination procedures"

The shape and topography of the site and the existing
physical facilities, such as buildings, pits, tanks, stacks
of barrels, fences, roads, overhead power lines, ponds,
etc., that are located on-site or adjacent to the site
should be identified and mapped as part of the initial site
investigation. (See chapter IV, Information Gathering.)
    Since no two sites are the same, this information is
vital for planning the best physical arrangement for
activities and functions at each particular site.
Similarly, information on the types and locations of
existing and potential hazards should be known so that a
safe layout can be accomplished. This information should be
included on a site map and used as the basis for developing
the master control plan.
    In addition to identifying the on-site location of all
waste piles, physical barriers, hazards, and special
problems on the site map, working areas must also be
identified. Such areas are needed for sampling, staging,
detoxifying, storing, processing, bulking, decontaminating,
treating, loading, transporting, and numerous support
functions. Some of these activities, like detoxification
and staging can take place in areas that are not free of
contamination. Other activities, like storing containers of
decontaminated materials or most support functions, must
take place in areas that are clean and free from
contamination. All such areas must be identified and
considered in the site layout plan.
    One key to a successful site operation is total control
of the entrance and exit of materials, equipment, and
personnel. The site map should therefore show the location
of existing and potential routes for entering and leaving
the site, especially emergency access routes. Separate

160

routes for personnel and equipment must sometimes be included. The site map is a very important tool to be used both in planning and in executing the control plan. It is a visual tool to compliment written plans or orders and must be updated as changes occur. Clear plastic overlays can be used to easily make changes and show current information on all posted maps.

The site map must be easily understood by all personnel and visitors if it is to be of maximum value. Every effort should be made to assure that the information contained on the map is both current and accurate.

One recommended method to prevent or reduce the transfer of contaminants and maintain control is to delineate zones or specific areas on the site where prescribed operations occur. The site must then be operated to insure that only those operations which are prescribed occur within a designated zone. Control of access points for entrance and exit to each of the zones or specific areas is the key to site-control as was discussed previously in this chapter. Movement of personnel and equipment between zones and onto the site itself are then limited to the access points. This should help keep contaminants within specified areas on-site and thus reduce the potential for spreading contamination. Three zones or control areas are usually designated for each site [2], (see Figure 1).

* Zone 1: Exclusion Zone
* Zone 2: Contamination Reduction Zone
* Zone 3: Support Zone

The use of a three-zone system, access control points, and exacting contamination reduction procedures provides a reasonable assurance against the translocation of contaminating substances and is highly recommended. A description of the purpose and layout of each zone follows.

Zone 1: Exclusion Zone

The Exclusion Zone is laid out to include all of the areas on-site where contamination is known or suspected to occur, and all of the areas where the processing of hazardous wastes is planned. The boundaries are initially established from information obtained during preliminary site investigations. Such information should include site records, visual observations, and instrument readings indicating the presence of possible contaminants. Organic or inorganic vapors or gases, harmful particulates in air,

161

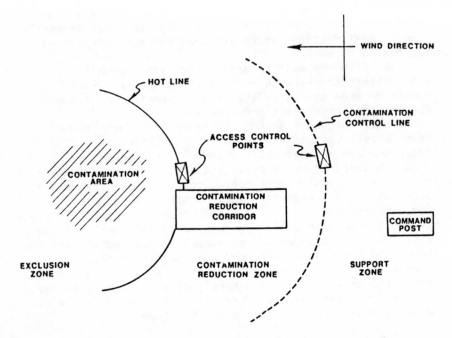

Figure 1.   Diagram of site work zones (no scale).

combustible gases, radiation, and the results of water and soil samples are used as indicators of the presence of possible contaminants. Other factors to consider when locating Exclusion Zone boundaries include the distance needed to prevent fire or explosion from affecting personnel and equipment outside the zone, the physical area necessary to conduct the various operations which directly involve the hazardous materials, and the potential for contaminants to be wind-blown from the area. (For hazardous materials incidents, the boundary of the Exclusion Zone is established by surveying the immediate environs of the incident to determine where the hazardous substances are located, noting where leachate or discoloration are visible, and determining drainage patterns.)

Once the boundaries have been determined, they should be physically secured and well-defined by visible landmarks. It is recommended that physical barriers such as fences, earth-berms, ditches, or barricades be erected around this zone to designate its location and to help control access to and from it. Signs must be posted and the use of bright-colored flagging or other visual material to draw attention to its location are very helpful. During subsequent operations, the location of zone boundaries may be moved or modified as required to meet program, operational, or environmental changes. Zone boundaries must be made to serve the purposes of the remedial actions taking place on-site.

When there is sufficient distance between the Exclusion Zone boundary and the work areas within the zone to protect entering personnel from unexpected venting of materials such as during a fire or explosion, the "hotline" or warning line for this zone should be established along the outer boundary of the zone. Otherwise, the "hotline" should be established up-wind of operations and separated from the Exclusion Zone by sufficient distance to account for such unexpected occurrences. When the "hotline" is not along the Exclusion Zone boundary, it must also be marked so as to be readily recognized.

As the name implies, all personnel and equipment are to be excluded from this zone unless they have specific permission of the on-site coordinator or site manager and are properly protected by the prescribed level of personal protective equipment. Prescribed levels are based on site-specific conditions which include the type of work to be done, the hazards that might be encountered, the physical condition of the worker, and environmental conditions such as weather. An entry and exit control point must be established at the periphery of the Exclusion Zone to regulate the flow of personnel and equipment into and out of the zone and to verify that the procedures established to enter and exit are followed.

Because many different activities can occur at the same time within the Exclusion Zone, different levels of protection are often justified. For example, the task of collecting samples from open containers might require one level of protection, while a walk-through air monitoring task or an observer task may require a different level of protection. The level of protection is determined by the measured concentration of substances in the work area, the potential for contamination while performing the task, the known or suspected presence of highly toxic substances, and perceived hazards--chemical and physical--of the site.

It is important that levels of protection be established commensurate with the actual or perceived hazards and not based on worst case scenarios that may not pertain to that particular site or site conditions. It may seem easier to designate one level of protection for everyone rather than worry about controlling several different levels in one zone. However, the assignment, when appropriate, of different levels of protection within the Exclusion Zone generally makes for a more flexible, effective, and less costly operation, while still maintaining a high degree of safety. Where different levels are permitted, areas within the zone must be conspicuously marked to identify the limits of each area and clearly labeled as to what levels of personal protective equipment are required for each area.

Hazardous wastes at uncontrolled sites are either treated and disposed of on-site or are prepared for safe shipment to an approved disposal site. In either case, the work of opening, sampling, emptying, bulking, mixing, detoxifying, treating, solidifying, filling, staging, and associated handling of hazardous materials, is done within the Exclusion Zone. Liquid and solid residues from the decontamination processes located in the Contamination Reduction Zone, and any other contaminated material from the site, are brought to the Exclusion Zone for treatment and disposal. If the wastes are to be properly confined so as not to spread contaminants, then all such functions must be controlled and contained within this zone.

The Exclusion Zone must be laid out to allow multiple operations (see Figure 2). Some remedial functions will occur simultaneously and others will occur sequentially. For example, there may be drum sampling, drum moving, drum staging and monitor well drilling all going on simultaneously, while bulking, solidifying, and loading would all be preceded by other operations. Site layout must be planned to assist in accomplishing the goals and sequences of the remedial actions which will be required at each particular site.

Figure 2.　An exclusion zone area showing
multiple operations.

During remedial actions in the Exclusion Zone, adequate
space must be provided for: (This is not intended to be an
all inclusive list.)

* Sampling and identifying wastes, including remote
  drum opening operations.
* Moving wastes. There must be room to maneuver waste
  handling equipment without bumping or spilling other
  wastes.
* Storing wastes. Wastes must be stored only with
  other wastes which are compatible. This may require
  several separate storage areas. Wastes are stored
  until they are treated an/or removed.
* Bulking wastes. Emptying the contents of drums or
  small containers into tanks or other large containers
  for processing or removal. There may be several
  bulking areas depending on waste compatibility and
  the volume of operations.
* Treating wastes. There must be room for each of the
  treatment processes which are to be used on-site.
  Some processes may have much larger space
  requirements than others, such as ponds, mixing
  basins, incinerators, reactor vessels, explosion
  pits, etc.
* A space buffer in case of fire or explosion
* Storing and treating contaminated run-off or surface
  water.
* Entrance and exit corridors for both equipment and
  personnel.
* Demolition of structures, tanks, or equipment and
  storage of the residue until their treatment and/or
  removal.
* Excavation of buried wastes, including room for the
  excavation to take place and storing both the
  excavated soil and wastes.
* Equipment storage for loaders, backhoes, drum
  grapplers, trucks, pumps, hoses, etc.
* Fire fighting or other emergency equipment to operate.
* Well drilling equipment to drill monitoring wells or
  test holes.

Zone 2: Contamination Reduction Zone

The Contamination Reduction Zone is laid out to
surround the Exclusion Zone and provide a buffer or
isolation area to separate contaminated areas from
uncontaminated or clean areas. This zone provides a

166

transition area to assure that the physical transfer of contaminating substances on personnel and equipment does not occur, or is limited to acceptable levels. This is accomplished by a combination of factors including control of access, decontamination procedures, distance between zones, work functions, and zone restrictions.

If all on-site contamination is contained within the Exclusion Zone boundaries, then the Contamination Reduction Zone will begin and remain a noncontaminated area. As operations proceed, the Contamination Reduction Zone will remain free from contamination only if it is properly maintained and managed. Some minor contamination may occur, but on a relative basis, the amount of contaminants should decrease from the "hotline" and not be detected at the Support Zone boundary. Any contamination that does occur should be removed and returned to the Exclusion Zone for treatment.

At the boundary between the Exclusion and the Contamination Reduction Zones, decontamination stations are established for personnel and equipment as required. The decontamination stations serve as control points to contain the contamination within the Exclusion Zone. Chapter 11, Decontamination, outlines requirements and procedures and gives necessary details on setting up and operating decontamination stations.

Access to the Contamination Reduction Zone from the Support Zone must also be controlled. Personnel and equipment must be allowed to enter only through designated control points and only when authorized. Personnel entering the Contamination Reduction Zone from the Support Zone must wear the personal protective equipment prescribed for this zone. The decontamination area always requires some level of protection. They should leave any contaminated protective equipment at the decontamination station when leaving the Contamination Reduction Zone. Personnel and equipment entering this zone from the Exclusion Zone will go through the decontamination stations and will leave all contaminated equipment and clothing at the stations. They must still maintain the specified level of protection required for this zone while passing to the Support Zone.

This means that the control or safety plan must address all activities and functions that take place at a remedial site and prescribe the personal protective equipment that will be used for each activity or function. The planning and execution of the plan to assure compliance are important steps to insure that the operation will be conducted safely commensurate with the actual problems of each function or activity.

Zone 3: Support Zone

The Support Zone is located in clean or noncontaminated areas outside of the Contamination Reduction Zone (see Figure 1), usually in the outermost portions of the site and sometimes in areas separated from the site. The Support Zone is generally located within an established site security perimeter, however some support activities such as personal vehicle storage and some emergency care facilities may be located outside of the security perimeter. Since support functions may be located in several different parts of the site, security must be maintained at required areas of the Support Zone.

Support functions such as the command post, equipment trailer, laboratory facilities, equipment storage, first aid station, training or briefing rooms, observation tower or platform, etc., are normally located within this zone. This does not mean that observation facilities, laboratory facilities, equipment storage, or first aid facilities cannot or are not located elsewhere, but only indicates that the Support Zone is set up to provide support functions in a clean area. Care must be exercised in locating each and every facility on-site. Site inspections have revealed that the level of contamination, based on air monitoring results, was greater in one support laboratory than was measured in the Exclusion Zone [3]. This, of course, may not always be the case, but it points out that planning and control are necessary in locating all activities on-site if a safe and efficient operation is to be maintained. Laboratories and other facilities that handle grossly contaminated samples or conduct tests that could release airborne contaminants, should be located within the Contamination Reduction Zone or the Exclusion Zone.

Personal protective equipment is not usually required within this zone. Normal work clothes are appropriate. Emergency respiratory protective equipment should be available in case of an explosion, fire, or unexpected occurrence, but this would be for an emergency action and not for normal operations. Contaminated personal clothing, equipment, and samples are not permitted in this zone, but are left in or taken to the decontamination stations in the Contamination Reduction Zone.

The location of the command post and other support facilities in the Support Zone depends on a number of factors. They include:

* Accessibility, topography, open space available, location of highways and railroad tracks, or other limitations.

168

* Wind direction, preferably the support facilities
  should be located up-wind of the Exclusion Zone.
  Shifts in wind direction and other conditions may
  dictate that the initial selection of certain support
  functions was not correct and they may have to be
  moved to a greater distance from work areas than was
  originally anticipated.
* Resources such as telephone and power lines, water,
  adequate roads for moving materials and equipment, and
  shelter.

Other functions that may be located in the Support Zone
include the communications area, the staging area for
supplies and equipment, a wind direction indicator visible
to all, a visitor briefing area or facility traffic control
and site security control headquarters, nurse station, and
other auxiliary functions that support a complex operation.

Some operations such as those provided by local police
and fire departments will not be located within the Support
Zone, but still must be considered in site layout. For
example, sufficient room must be provided for access and
maneuverability of special equipment, and water sources must
be identified for fire fighting operations. Access routes
must be provided and kept unobstructed for ambulances or
emergency vehicles.

The location and operation of all functions at such
sites is dictated largely by what the function does or is
supposed to do. If it is a clean operation or must be kept
clean, it should be located in the Support Zone. If it is
an operation that involves sampling or handling of
contaminated materials it should be located in either the
Exclusion Zone or Contamination Reduction Zone.

AREA DIMENSIONS

The distance between various on-site functions must all
be site-specific. The size of each zone or work area must
also be based on the conditions at each site and not on some
standard formula. Considerable judgment is needed to assure
that the distances between zone boundaries and distances
between activities within zones are large enough to provide
room for the necessary operations. The distances must also
be great enough to prevent the spread of contaminants and
eliminate or significantly reduce the possibility of injury
due to explosion or fire.

The following criteria should be considered in
establishing area dimensions and boundary distances:

* Physical and topographical features of the site,
  including the exterior dimensions of the site itself.
* Weather conditions.
* Field and laboratory measurements of air contaminants
  and environmental samples.
* Air dispersion calculations.
* Potential for explosion and flying debris.
* Physical, chemical, toxicological, and other
  characteristics of the substances present.
* Clean-up activities required.
* Potential for fire both on-site and to surrounding
  areas.
* Area needed to conduct operations.
* Decontamination procedures
* Dimensions of contaminated areas.
* Potential for exposure.

Boundaries should not be considered permanent but should be thought of as flexible and moveable as the demands of the site change.

CONTAMINATION CONTROL

The project manager, team leader, or on-site coordinator is responsible for control of all on-site activities. The safety officer only assists, but should have the authority to stop an operation if it is found to be life-threatening. A site will only be as safe as those who are in control are willing to make it, and contamination will only be contained if control measures are planned and implemented.

Procedures must be established to assure that the clean zone remains clean. This is done through utilizing proper clean-up and mitigation methods in the Exclusion Zone and proper decontamination methods in the Contamination Reduction Zone. It is also done by controlling entrances and exits of all zones as indicated earlier, and in providing the facilities and separation necessary to safely accomplish the work.

An elevated observation tower or strategically located video cameras provide useful methods to observe safety procedures, work progress, and work procedures. During an emergency, having visual observation capabilities can be very helpful. This method can also be used to observe ways to improve existing procedures, and visitors can be given a visual tour of the site without having to enter restricted areas or be trained in the use of personal protective equipment.

To verify that site control procedures and existing boundary locations are preventing the spread of contamination, a monitoring and sampling program should be established. Zones should be regularly monitored to provide the necessary data from which to make management decisions. Operations should involve reasonable methods to determine if material is being transferred between zones, to assist in modifying zone or site boundaries, and to modify site safety or control plans as required. All such monitoring must be kept within the capacity of the analytical support functions available or it will be of limited or no value in making rapid decisions. Where applicable, direct reading instruments can be used to monitor vapor, gases, and particulates in the air. Some data on contaminants in water can be obtained rapidly but normally most analyses for water and soil take several days to obtain results.

The use of a three-zone system, access control points, effective clean-up, and exacting decontamination procedures provide a reasonable assurance that contaminating substances can be controlled. Much of the information presented in this chapter is based on a "worst case" situation. Less stringent site control and decontamination procedures may be utilized if more definitive information is available on the types of substances involved and the hazards present. The necessary information can be obtained through air monitoring, instrument surveys and sampling, and technical data concerning the characteristics and behavior of the material present. Experience gained from previous remedial work is very valuable in assessing the conditions and dangers associated with this type of work. Most decisions are judgmental and rely on the ability and experience of the decision-maker.

Each site is different, hazards are different, conditions are different, personnel are often different, and the regulatory agency is sometimes different. The site layout procedures given in this chapter must be adapted to the unique situation at each individual site.

ENGINEERED CONTROLS

Many hazardous or potentially hazardous conditions on-site can be eliminated or significantly reduced by employing engineered controls. Unsafe conditions, such as rutted or bumpy roads for forklifts carrying hazardous chemicals, sharp objects that could rip protective clothing, inadequate warning devices to warn of backing equipment, fire and explosion hazards, inadequate illumination in closed areas or at night, excess noise levels, confined spaces without proper ventilation, and buried or overhead electrical cables are all examples of potential causes of accidents or injuries at hazardous waste sites. Effective engineering can help prevent such accidents. Remote drum handling or drum opening devices, protective berms or embankments, sparkless tools, vacuum pumps and tanks, overpack drums, and controlled drainage areas are examples of engineered controls that can be used to isolate or control hazardous materials handling activities and thus reduce human exposure to hazardous substances.

Occupational safety and health professionals consider the use of engineered safeguards that contain, isolate, or remove the hazard from the worker to be superior to the use of personal protective equipment that attempts to isolate the worker from the hazard. This approach works well in industrial settings where ventilation systems and physical barriers are employed to isolate or remove the hazard and leave the worker in a safe environment. This approach is considerably more difficult to accomplish at hazardous waste sites because of the outdoor conditions that normally exist at such sites and because the operations are usually of much shorter duration than in industrial settings. Nevertheless, it is still a better policy to control the hazard by isolation, processing, containment, and removal in conjunction with proper worker protection than to rely solely on protecting the worker. When the hazards are controlled, all personnel are protected. If personal protective equipment is the only protection provided, it requires independent effort by each worker involved, and may not be as effective.

Engineered safeguards to isolate hazards from workers can include remote handling devices such as hydraulic drum handlers (see Figure 3), sparkless drum openers, pneumatic bung openers for drums, explosion or fire pits, berms of earth to form barriers between wastes and workers, explosion shields, explosion containers, and drum overpacks.

Selection of alternate processes or activities can also provide an engineered safeguard for personnel. If the suggested treatment procedure for a particular waste was incineration and there was a danger of explosion of a waste stored nearby, an alternate process of chemical mixing, solidifying with inert materials, and containerizing might provide a solution with less risk to personnel. All such alternate approaches should be considered during planning and operation to assure that the best possible alternative is used commensurate with the problem.

Engineered safeguards, alternate processes, special equipment, and engineered construction techniques can and should be used whenever possible at a hazardous waste site.

Figure 3. Hydraulic drum handler (drum grappler).

Before remedial actions are begun, the area can be prepared to facilitate operations and eliminate obvious physical hazards. Roadways, work areas and storage areas can be constructed to provide ease of access and a sound roadbed for heavy equipment and vehicles. Traffic control patterns can be selected or modified to facilitate efficiency of operations. Security fences or barricades can be erected. Communication facilities can be installed. Work areas can be cleared and physical hazards can be eliminated as much as possible. Some physical hazards to consider include:

* Ignition sources in flammable hard areas such as drum opening and bulking areas.
* Exposed and/or underground electrical wiring and low overhead wires which may be cut or entangled in equipment resulting in electrical shocks, short circuits, and possible fires.
* Sharp protruding edges such as torn metal, glass, nails, and other objects which can puncture or tear protective clothing or equipment.
* Protruding objects which can cause slips, trips, and falls.
* Unsecured railings, loose steps or flooring, holes, slippery surfaces, debris, and other obstacles that can cause slips, trips, and falls.
* Weeds and debris which obstruct visibility. Unexpected hazards may be hidden from view by such obstructions and impaired visibility can result in accidents during handling and transportation of wastes and containers.

Weeds and debris can be removed, walking surfaces can be cleared and repaired, skid resistant strips can be installed on slippery surfaces, railings can be repaired or installed, stairs and ladders can be secured, electrical wiring can be repaired or relocated, and sharp objects and protruding edges which cannot be removed can be covered or properly guarded. Operation pads for mobile facilities and temporary structures can be constructed. Loading docks, wash-down process facilities, water treatment facilities, and staging areas can be constructed to facilitate safe and effective operations.

ISOLATION AND CONTAINMENT BARRIERS

Containment of materials can often be achieved by the construction of berms and dikes of suitable on-site soils, or in rare cases, imported soils. Liners may be required to contain the liquids if porous soils are used. Liners can also help to reduce the amount of contaminated soil that may require disposal. Such barriers will contain spills and releases of liquids or solids; however, splashing liquids, liquids under pressure, and gases may not be adequately contained. Trenches can be constructed to collect or divert spills and releases of liquids or contaminated water for later treatment and disposal, e.g., leachate and contaminated runoff. Dispersion of wastes to both on-site and off-site areas can thereby be controlled.

Access to contaminated areas or potential areas of contamination can be controlled by exclusion barriers such as fences, trenches, or earth-berms to keep people and vehicles out, thereby preventing inadvertent entry into hazardous areas. Signs are required by law to be posted at hazardous waste facilities. Streamers, flags, ribbons, or other visual warning can alert persons to hazard areas and prevent accidental entry into such areas.

SPECIAL EQUIPMENT

Equipment can be selected, developed, or modified to reduce direct contact of workers with wastes and containers during handling and transport of materials (see Figure 4). Containers can be moved by machines such as Bobcats, forklifts, drum grapplers, backhoes, and cranes. Moving and loading of containers, bulk solids and small spills can be achieved by front-end loaders, backhoes, draglines, Bobcats., etc. Splash plates or shields and cab enclosures can prevent or reduce worker exposure to spills, splashes, and releases under force, e.g., liquids and solids released under pressure, explosions, forceful rupture of containers, etc. In extreme cases, total cab enclosures with temperature controls and self-contained air supplies could be necessary to provide total protection from vapors, solids, and liquids while operating in hazardous areas.

Special equipment such as reactor vessels for on-site chemical treatment, tanks or ponds for wastewater treatment, vacuum pumps and vessels, pneumatic pumps, bulking and handling equipment, and similar items should be reviewed by qualified engineers and safety and health personnel for

Figure 4.   Special equipment used at remedial sites
(drum punch, bulking chamber).

technical suitability and safety. Workers using the
equipment and other personnel performing activities in the
same area must be protected from improperly designed or
constructed special equipment.

WARNING ALARMS AND DEVICES

Emergency alarms should be provided to alert personnel
of emergency conditions such as fire, dangerous weather
conditions, accidents, material releases, etc. All
personnel should be familiar with the different alarms,
signals, and warnings established for different types of
emergencies.

For operations where vehicles or equipment must backup,
and for loaders carrying containers in front which partially
block visibility, the use of warning devices is advisable
and in most cases required. Flashing lights and audible
warning devices such as beepers or horns are necessary to
alert workers who may not be watching or whose vision and
hearing may be reduced because of their personal protective
equipment.

Certain hazardous conditions such as vapor and oxygen
concentrations can be monitored by devices which will give
an audible alarm when concentrations reach a pre-set level.
Use of these devices in enclosed spaces is particularly
advisable. Heat and smoke detectors can be used to alert
workers and off-site response groups to fires and potential
fire conditions. Such devices should be used whenever
workers are involved in activities adjacent to the potential
hazardous conditions. Heat and smoke detectors can signal
fire departments and trigger alarms when personnel are not
working such as evenings and weekends. For further
discussion of emergencies, see chapter XIII, Contingency
Planning.

AIR POLLUTION CONTROL

Control or minimization of the concentrations of air
contaminants such as mists, vapors, and particulates can be
achieved several ways. Using devices to collect or draw off
the contaminants from the work area, application of water or
other appropriate liquids, encapsulation of the wastes,
dispersion and dilution of contaminants, proper handling of
wastes and debris, permissible controlled releases at

specified intervals or under controlled conditions, and selection of alternative methods which reduce generation of air contaminants.

Gas collection systems such as passive gas trenches, passive trench barriers, and active gas extraction wells [4] are designed to control gas migration through soils and subsequent releases to the air or to other areas containing buried hazardous wastes. If necessary, filters, flares, or other methods can be used to treat gases vented from these types of systems.

Ventilation systems consist of air moving devices, hoods or vents, ductwork, and air cleaning devices (when appropriate). Such systems can either collect the contaminants at the source using canopy hoods or enclosures (i.e., local ventilation), or dilute the level of contaminants by exhausting the air through appropriately placed vents (i.e., general ventilation). The air flow is usually supplied for these systems by intrinsically safe fans. Fans can be located to either draw air into the system by negative pressure or force air into the system by positive pressure. General ventilation basically reduces or dilutes contaminants while local ventilation is designed to prevent any exposures to workers.

Mobile on-site laboratories are equipped with hoods to prevent exposure to lab workers. Most sites are not conducive to the application of large ventilation systems. If ventilation systems are required or if they can be cost-effective, they should be designed by qualified safety engineers or industrial hygienists.

Exhausts from ventilation systems can either be released (based on proper dispersion models) or treated by applicable pollution control devices such as electrostatic precipitators, bag house systems, etc. This decision should be based on the concentration of contaminants in the exhaust stream and the potential hazards of each pollutant. Treatment residues must be disposed of in accordance with applicable regulations. Any major use of ventilation systems for contaminant control should carefully consider the safety, effectiveness, practicality and costs of such systems.

The use of general ventilation by forced air to aid in the dispersion of contaminants in the working area, to aid in worker comfort in hot weather, or to add warmth during cold weather should be carefully monitored and controlled so as not to disperse contaminants to other locations on-site or off-site. Properly located exhaust stacks can make good use of normal dispersion for dilution of contaminants providing exhaust levels do not exceed acceptable limits [5].

Wetting down road surfaces serves to reduce the amount of dust generated and assists in compaction of the road surfaces [6]. Dust controls should be employed using water sprays and other suitable liquids. Contaminated road dusts can be reduced by wise use of wetting agents and liquids, applied by water wagons or similar mobile applicators.

Liquid sprays and curtains can be used to reduce dust during transfer, handling, and packaging of bulk wastes and wastes with particle sizes capable of air dispersion such as powders and finely ground compounds. For some wastes, this may require the use of special wetting agents to aid in adherence or adsorption to the particles. The wetted materials must be collected, contained, and disposed of in accordance with applicable regulations to prevent redispersion [7]. Similar wetting of waste piles, working areas, and containers can serve to reduce dust during materials handling operations and clean-up activities as long as the wetting agents are compatible with the wastes. Care must also be taken to avoid using too much water and spreading pollutants by the run-off.

Reduction of airborne contaminants can also be achieved by the clean-up of debris and containment of wastes. This can be achieved by using appropriate cover materials such as tarps, caps or sealants, and containers such as drums, overpacks, covered tanks, and dumpsters with closable lids. This normally involves common sense and good housekeeping practices. However, the selection of appropriate materials for caps and sealants is a technical decision.

Surface and reburial can be used to control hazardous gas and dust emissions for extended periods of time or as permanent remedial actions [4,8]. Short-term or temporary controls such as the use of tarps and containers mentioned above, are more applicable to clean-up operations involving off-site disposal. Contaminated tarps must be disposed of by incineration or burial if they cannot be decontaminated.

The cost of providing control of air pollutants and protecting worker health and safety should be a major consideration in the selection of acceptable alternative remedial actions. Unfortunately, it is often difficult to obtain accurate estimates for increased costs associated with protecting workers' safety and health [9]. Efforts should be made to select methods and procedures for tasks at remedial action sites that reduce exposures by minimizing generation of dusts, vapors, and mists.

CONTAMINATED SURFACE WATER CONTROL

Exposure of workers to contaminated water and migration of contaminated water to clean work areas or off-site areas

can be prevented or minimized by diverting rain run-off away
from wastes, and collecting storing and treating the
contaminated water that is generated. Dikes, berms,
drainage pipes, ditches, and contour grading are some of the
engineered controls which are used to keep surface water
from flowing onto hazardous waste locations. Similar
devices or controls are used to direct contaminated run-off
water coming from the site to collection basins and ponds
for pre-treatment and/or transfer to on-site treatment
facilities, or to off-site treatment and disposal facilities
[4,8]. In areas of insufficient water for fire protection,
rain or surface water that is kept from entering the site is
a possible source for this purpose. Run-off water or
leachate from the site should not be used for fire control.

REFERENCES

1. "Personal Protection and Safety-Training Manual,"
    Hazardous Response Support DIvision, National Training
    and Technology Center, U.S. EPA, Cincinnati, OH, 1982.
2. "Interim Standard Operating Safety Guides," Office of
    Emergency and Remedial Response, U. S. EPA, Edison, NJ,
    1982.
3. Costello, R. J., C. Geraci, P. Eller, and R. Ronk.
    "Health Hazard Evaluation Determination Report IA
    82-40, Triangle Chemical Site, Bridge City, Texas",
    National Institute for Occupational Safety and Health,
    Cincinnati, OH, 1983.
4. Walsh, J. J., and D. P. Gillespie, "Selecting Among
    Alternative Remedial Actions for Uncontrolled Hazardous
    Waste Sites", U. S. EPA, Cincinnati, OH, 1982.
5. "Documentation of Threshold Limit Values", 4the Rev.
    Ed. - 1980, with 1982 changes and editions, ISBN 0-9367
    12-12-9, American Conference of Governmental Industrial
    Hygienists, Cincinnati, OH, 1982.
6. Church, H. K., Excavation Handbook (McGray - Hill Book
    Co., New York, 1981.
7. Fundamentals of Industrial Hygine (National Safety
    Council [NSC], Chicago, 1971.)
8. Rogeshewski, P., H. Bryson, and K. Wagner, "Handbook
    for Remedial Action at Waste Disposal Sites", U. S. EPA
    Report 625/6-82-006, Cincinnati, OH, 1982.
9. Lippitt, J.M., J. Walsh and A. D. Puccio, "Costs of
    Remedial Actions at Uncontrolled Hazardous Waste
    Sites: Worker Health and Safety Considerations," in
    Proceedings of Conference on Hazardous Wastes and
    Environmental Emergencies, Houston, TX, March 12-14,
    1984.

PERSONAL PROTECTIVE EQUIPMENT

Arthur D. Schwope
E. Robinson Hoyle
   Arthur D. Little, Inc.
   Acorn Park
   Cambridge, Massachusetts 02140

A variety of known and unknown chemical and physical health hazards are potentially present on a hazardous waste site. Consequently, remedial action personnel must be properly protected in order to prevent short- and long-term disabilities related to these hazards. Personal protective equipment (PPE) and administrative controls (e.g., reducing an individual's exposure time by personnel rotation) are the primary means for achieving this objective. This chapter will address the state of the art PPE which is available, the combinations or ensembles of PPE suited to the potential hazards, and the effectiveness of protective clothing as barriers to chemicals. The focus is on chemical rather than physical hazards. Hence, the two principal topics are respirators and chemically resistant clothing.

TYPES OF PPE

A large variety of chemicals and chemical mixtures potentially present at a site and can range from benign to extremely toxic poisons and carcinogens. (See chapter 6.) Corrosive, explosive, flammable and radioactive materials, as well as oxygen deficient confined spaces may also be present [1]. Although uncommon, potentially hazardous biologically active materials such as bacteria and viruses may also be present. Physical stresses on-site may include high noise (excavation machinery, drum crushing) and heat or cold stress (heat stress from confinement in PPE).

Finally, uneven terrain, vegetation, sharp metal and glass, and equipment malfunctions are also potential causes of injury on a site [2].

Such hazards can produce a multitude of different health effects, including:

  *temporary or permanent damage to the eyes, ears, skin, internal organs, or the nervous or circulatory systems
  *carcinogenicity, mutagenicity or teratogenicity
  *loss of limbs, organs, or death

Chemical effects on health are dependent on the route and duration of exposure and the concentration and toxicity of the chemical. For any given exposure the effects may also vary from individual to individual. The route of exposure may be by ingestion of solids and liquids, skin absorption of liquids and gases, and inhalation of gases [3,4].

As stated earlier, the primary exposure control method on a hazardous waste site is personal protective equipment. The types of PPE available include full body clothing, respiratory equipment, and head, hand, eye, ear, face, and foot protection. Hard hats and safety shoes or boots are typically used to protect the head and feet from falling or flying objects. Hard hats can be equipped with ear muffs, face shields, and goggles and are available in different sizes, shapes, and materials. Safety shoes or boots can be made of leather, plastic, rubber or combinations thereof, and are equipped with a metal toe box and a steel shank along the sole of the shoe. Since leather readily absorbs most liquids, plastic and rubber footwear are used for protection from water and other liquids which may contain toxic chemicals. In the later case it is important to select footwear fabricated from materials which resist absorption of the toxic chemicals. This will be discussed later in this chapter. Hard hats and safety footwear should be properly fitted to the individual and selected to provide the necessary degree of protection [5].

Eye, ear, and face PPE can include safety glasses or goggles, ear muffs, ear plugs, and faceshields. Tinted safety glasses and goggles are available for welding and burning applications. Goggles can be gas-tight to prevent gases and vapors from contacting the eyes or vented to allow for good ventilation, depending on the application. Faceshields protect the entire face and in some cases the neck region. Good fitting eye, ear, and face protection is critical to providing protection as well as not introducing additional hazards. Equipment from several manufacturers may be required in order to assure that every worker has a good fit for each item. Poorly fitting ear plugs will not attenuate the noise, and loosely fitting safety glasses or goggles may create an additional hazard during strenuous fast-paced work. The advantages and limitations of these protective devices should be fully defined and understood by the individual user [2,5].

Hoods used in combination with hard hat, goggles, and respirators will adequately cover the back of the neck and head from spills and splashes of chemicals. Aprons and sleeves (disposable and non-disposable) may be utilized, but there are limitations in regard to their durability and coverage.

Full-body PPE or splash suits and the associated splash protection clothing (gloves, booties, hoods, etc.) can be divided into three categories: one-piece suits, fully-encapsulating suits and multiple-piece full-body suits with hoods and boots. This latter group includes pants and suits made of materials such as neoprene rubber, butyl rubber, polyvinyl chloride (PVC), and polyethylene. One-piece disposable garments with hoods and booties are also available, and are typically fabricated from polyethylene, polypropylene or PVC [2,6].

Encapsulating suits provide a self-contained internal environment for the wearer. All parts of the body, including head, hands, and feet are enclosed. These suits must be equipped with a supplied-air system to provide the user with a fresh air source. Fully-encapsulating suits can be chemical resistant, as well as heat-resistant, thus permitting personnel to work in dual-hazard environments. In addition to protecting the body, single- and multi-component suits prevent contamination of the individual's street and underclothing [7]. This reduces the potential for spreading the hazard to lunch and smoking areas, offices, automobiles and homes.

Gloves and respirators are considered the key items of personal protective equipment. Gloves should permit an individual to handle equipment and materials while providing a highly resistant barrier between hazardous chemicals and the skin. The effectiveness of gloves to provide a barrier has become an important issue in recent years. Issues of chemical resistance and permeability of glove materials and other clothing materials such as butyl rubber, neoprene, and natural rubber, etc., are discussed later in this chapter.

RESPIRATORS

The purpose of respiratory equipment on a hazardous waste site is to eliminate a worker's inhalation of hazardous gases and vapors. The respiratory protection selected for this purpose must be chosen according to the particular chemical or group of chemicals that may be encountered. Respiratory protective equipment is divided into two groups: air-purifying and supplied-air respirators [7].

Air-Purifying

Air-purifying respirators utilize a filter to remove any contaminants which are inhaled. There are a number of different filters or cartridges to filter particular chemicals or chemical groups (see Table I) [8,9].

Air-purifying respirators are available in different facemask configurations depending on the degree of protection which is required. The quarter face mask covers the nose and mouth and rests on the bridge of the nose and front of the chin. The half face mask provides protection for the nose and mouth but fits underneath the chin. The full face mask fits around the outer perimeter of the face covering the eyes, nose, and mouth. The full-face mask can have single or double filters or the cartridge cannister can be connected to the mask by a hose (e.g., gas mask with cannister).

Each of the air-purifying masks provide different degrees of protection from contaminants. Because of these differences, each of the facepieces have been assigned protection factors (PF), as listed in Table II [7].

These protection factors are a measure of the overall effectiveness of a respirator including the filtering efficiency and fit of the face to facepiece seal as determined by quantitative fit testing. In cases where contaminant concentrations (with threshold limit values or permissible exposure limits) are known, these protection factors can be utilized to estimate the degree of protection the air-purifying mask can provide.

Another type of air-purifying respirator is the air-powered respirator or the personal engineering control device (PECD). The PECD is a helmet, hood, or mask equipped with a motor to pull air through a filter and push the filtered air into the mask or helmet. This feature gives a positive outflow of air from the mask. This positive outflow provides an added barrier to contaminants reaching the face as opposed to the other air purifying masks which require the user to inhale through one or more filters, thereby creating an inflow of potentially contaminated air to facial area. Due to the positive outflow or pressure, the air-powered air-purifying mask has a protection factor

Table I. Respirator Selection by Colors Assigned to
Atmospheric Contaminants

Atmospheric Contaminants to be protected against	Colors Assigned*
Acid gases---------------------	White
Hydrocyanic acid gas-----------	White with 1/2-inch green stripe completely around the canister near the bottom
Chlorine gas-------------------	White with 1/2-inch yellow stripe completely around the canister near the bottom
Organic vapors-----------------	Black
Ammonia gas--------------------	Green
Acid gases and ammonia gas----- stripe completely around the	Green with 1/2-inch white canister near the bottom
Carbon monoxide----------------	Blue
Acid gases and organic vapors--	Yellow
Hydrocyanic acid gas and chloropicrin vapor-----------	Yellow with 1/2-inch blue stripe completely around the canister near the bottom
Acid gases, organic vapors,---- and ammonia gases	Brown
Radioactive materials,--------- excepting tritium and noble gases	Purple (Magenta)
Particulates (dusts, fumes,---- mists, fogs, or smokes) in	Canister color for contaminant as designated combination with any of the above, with 1/2-inch gray above gases or vapors stripe completely around the canister near the top
All of the above atmospheric--- contaminants	Red with 1/2-inch gray stripe completely around the canister near the top

*Gray shall not be assigned as the main color for a canister
designed to remove acids or vapors.
NOTE: Orange is used as a complete body or stripe color to
represent gases not included in this table. The user will
need to refer to the canister label to determine the degree
of protection the canister will afford.

Table II. Respirator Protection Factors[1]

Type Respirator[2]	Facepiece Pressure	Protection Factor
I. Air-Purifying		
A. Particulate[3] removing		
Single-use,[4] dust[5]	−	5
Quarter-mask, dust[6]	−	5
Half-mask, dust[6]	−	10
Half- or Quarter-mask, fume[7]	−	10
Half- or Quarter-mask, High-Efficiency[8]	−	10
Full Facepiece, High Efficiency	−	50
Powered, High-Efficiency all enclosures	+	1000
Powered, dust or fume, all enclosures	+	x[9]
B. Gas and Vapor-Removing[10]		
Half-Mask	−	10
Full Facepiece	−	50
II. Atmosphere-Supplying		
A. Supplied Air		
Demand, Half-mask	−	10
Demand, Full Facepiece	−	50
Hose Mask Without Blower Full Facepiece	−	50
Pressure-Demand, Half Mask[11]	+	1000
Pressure-Demand, Full Facepiece[12]	+	2000
Hose Mask With Blower, Full Facepiece	−	50
Continuous Flow, Half-Mask[11]	+	1000
Continuous Flow, Full Facepiece[12]	+	2000
Continuous Flow, Hood	+	2000
Helmet, or Suit[13]	+	2000
B. Self-Contained Breathing Apparatus (SCBA)		
Open-Circuit, Demand, Full Facepiece	−	50
Open-Circuit, Pressure-demand Full Facepiece	+	10,000[14]
Closed-Circuit, Oxygen Tank-type, Full Facepiece	−	50

TABLE II. Respirator Protection Factors[1]--continued

Type Respirator[2]	Facepiece Pressure	Protection Factor
III. Combination Respirator		
A. Any combination of air-atmosphere-supplying respirator	Use minimum protection factor listed above for type of mode of operation	
B. Any combination of supplied-air respirator and an SCBA		

Exception: Combination supplied-air respirators, in pressure-demand or other positive pressure mode with an auxiliary self-contained air supply, and a full facepiece, should use the PF for pressure-demand SCBA.

NOTE: Table is not to be reproduced without the accompanying footnotes.

SOURCE: Pritchard, John A. (1977). A Guide to Industrial Respiratory Protection. U. S. Energy Research and Development Administration.

----

[1]The overall protection afforded by a given respirator design (and mode of operation) may be defined in terms of its protection factor (PF). The PF is a measure of the degree of protection afforded by a respirator, defined as the ratio of the concentration of contaminant in the ambient atmosphere to that inside the enclosure (usually inside the facepiece) under conditions of use. Respirators should be selected so that the concentration inhaled by the wearer will not exceed the appropriate limit. The recommended respirator PF's are selection and use guides, and should only be used when the employer has established a minimal acceptable respirator program as defined in Section 3 of the ANSI Z88.2-1969 Standard.

[2]In addition to facepieces, this includes any type of enclosure or covering of the wearer's breathing zone, such as supplied-air hoods helmets, or suits.

[3]Includes dusts, mists, and fumes only. Does not apply when gases or vapors are absorbed on particulates and may be volatilized or for particulates volatile at room temperature. Example: Coke oven emissions.

[4]Any single-use dust respirator (with or without valve) not specifically tested against a specified contaminant.

[5]Single-use dust respirators have been tested against asbestos and cotton dust and could be assigned a PF of 10 for these particulates.

[6]Dust filter refers to a dust respirator approved by the silica dust test, and includes all types of media, that is, both nondegradable mechanical type media and degradable resin-impregnated wool felt or combination wool-synthetic felt media.

[7]Fume filter refers to a fume respirator approved by the lead fume test. All types of media are included.

[8]High-efficiency filter refers to a high-efficiency particulate respirator. The filter must be at least 99.97% efficient against 0.3 um DOP to be approved.

[9]To be assigned, based on dust or fume filter efficiency for specific contaminant.

[10]For gases and vapors, a PF should only be assigned when published test data indicate the cartridge or canister has adequate sorbent efficiency and service life for a specific gas or vapor. In addition, the PF should not be applied in gas or vapor concentrations that are: 1) immediately dangerous to life, 2) above the lower explosive limits, and 3) cause eye irritation when using a half-mask.

[11]A positive pressure supplied-air respirator equipped with a half-mask facepiece may not be as stable on the face as a full facepiece. Therefore, the PF recommended is half that for a similar device equipped with a full facepiece.

[12]A positive pressure supplied-air respirator equipped with a full facepiece provides eye protection but is not approved for use in atmospheres immediately dangerous to life. It is recognized that the facepiece leakage, when a positive pressure is maintained, should be the same as an SCBA operated in the positive pressure mode. However, to emphasize that it basically is not for emergency use, the PF is limited to 2,000.

[13]The design of the supplied-air hood, suit, or helmet (with a minimum of 6 cfm of air) may determine its overall efficiency and protection. For example, when working with the arms over the head, some hoods draw the contaminant into the hood breathing zone. This may be overcome by wearing a

short hood under a coat or overalls. Other limitations specified by the approval agency must be considered before using in certain types of atmospheres.

[14]The SCBA operated in the positive pressure mode has been tested on a selected 31-man panel and the facepiece leakage recorded as less than 0.01% penetration. Therefore, a PF of 10,000+ is recommended. At this time, the lower limit of detection 0.01% does not warrant listing a higher number. A positive pressure SCBA for an unknown concentration is recommended. This is consistent with the 10,000+ that is listed. It is essential to have an emergency device for use in unknown concentrations. A combination supplied-air respirator in pressure-demand or other positive pressure mode, with auxiliary self-contained air supply is also recommended for use in unknown concentrations of contaminants immediately dangerous to life. Other limitations, such as skin absorption of HCN or tritium, must be considered.

of 1000 [7]. Air-powered respirators can be equipped with organic vapor filters, or particulate filters, or both kinds of filters.

An air-purifying respirator must be used in accordance with certain federal regulations. It should be approved by the National Institute for Occupational Health and Safety (NIOSH) and the Mine Safety and Health Administration (MSHA) [10]. Equipment which meets the requirements of NIOSH/MSHA are usually labeled as such. If approved, a label on the respirator will say so and will also provide information as to the type and degree of protection provided against dusts, fumes, mists, and radon daughters. The label also carries a warning that air-purifying devices cannot be used in atmospheres with less than 19.5 percent oxygen. The NIOSH approval label provides valuable information, such as limitations and cautions, and assures the performance of the equipment under NIOSH/MSHA testing conditions.

Supplied Air

The second type of respiratory equipment is the supplied-air respirator; this group of respirators can be further divided into supplied air breathing systems and self-contained breathing apparatus (SCBA) (Table III) [7]. Supplied air breathing systems incorporate a compressor or compressed air cylinder which provides the wearer's mask or hood with fresh air. The user is attached to the fresh air source by a hose and is, consequently, only semi-mobile. The critical feature of these systems is the fresh air source. Compressor inlets should be located and fitted with appropriate filters to prevent introduction of hazardous gases, particulates, or vapors to the air delivered to the user. Supplied air systems are used only minimally because of the difficulties in providing contaminant-free air and the semi-portable feature, which limits the mobility of the worker [11].

The SCBA is composed of a portable cylinder of pressurized fresh air carried on the user's back and associated breathing hose, air flow regulator, and full face mask. Typical cylinders have a 30-minute air supply. Both the SCBA and supplied-air breathing system operate on either pressure demand or demand air supply (Table III). Some systems can operate on both pressure demand and demand air supply. The pressure demand system supplies a positive outflow from the mask, reducing the potential for contaminant entry. In the demand system, the air is supplied to the mask as demanded or required by the user.

Table III. Self-Contained Breathing Apparatus

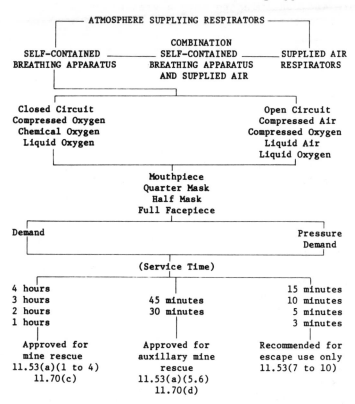

```
            ┌── ATMOSPHERE SUPPLYING RESPIRATORS ──┐
            │                                       │
                           COMBINATION
   SELF-CONTAINED ──────  SELF-CONTAINED ─────── SUPPLIED AIR
  BREATHING APPARATUS    BREATHING APPARATUS     RESPIRATORS
                          AND SUPPLIED AIR
        │                                             │
  Closed Circuit                              Open Circuit
  Compressed Oxygen                           Compressed Air
  Chemical Oxygen                             Compressed Oxygen
  Liquid Oxygen                               Liquid Air
        │                                     Liquid Oxygen
        └──────────────────┬──────────────────────┘
                     Mouthpiece
                     Quarter Mask
                     Half Mask
                     Full Facepiece
        ┌──────────────────┴──────────────────────┐
  Demand                                      Pressure
        │                                     Demand
        └──────────────────┬──────────────────────┘
                     (Service Time)
        ┌──────────────────┼──────────────────────┐
  4 hours                                    15 minutes
  3 hours              45 minutes            10 minutes
  2 hours              30 minutes            5 minutes
  1 hours                                    3 minutes
        │                  │                      │
  Approved for        Approved for         Recommended for
  mine rescue         auxillary mine       escape use only
  11.53(a)(1 to 4)    rescue               11.53(7 to 10)
     11.70(c)         11.53(a)(5.6)
                      11.70(d)
```

SOURCE: Pritchard, John A. (1977). _A Guide to Industrial Respiratory Protection._ U. S. Energy Research and Development Administration.

Consequently, in the demand mode the mask is under negative pressure and contaminants have a greater probability of leaking into the mask. Protection factors for these supplied-air systems range up to 10,000+ for those with full face masks operated in the pressure demand mode.

Fit

In order for a respirator to provide the intended level of protection, it must be fit to the individual's face and worn properly. There must be a good seal between the respirator and the face. Factors which can comprise the seal are:

*the limited number of standard respirator sizes for the almost infinite number of face sizes and shapes (e.g., hollow temples and very prominent cheekbones);
*facial hair;
*scars and; and
*lack of teeth or dentures.

Respirator fit can be assessed by quantitative and qualitative methods. Quantitative fit tests involve placing an individual wearing the respirator in a chamber containing a non-toxic gas, vapor, or aerosol (i.e., a tracer compound). The air inside the respirator is then monitored for the presence of the tracer by means of analytical instrumentation. Although very effective, quantitative methods are usually restricted to laboratory settings and require highly trained personnel.

Qualitative fit testing does not require sophisticated or expensive equipment, and can easily be performed in the field. Such testing typically comprises three procedures:

*negative pressure test;
*positive pressure test; and
*odor test.

In the negative pressure test, all air entry ways to the respirator are closed off. This can be accomplished by placing one's hand over the inlet of the canister, cartridge(s), or filters, or squeezing the breathing tube on an air-supplied respirator. The wearer than inhales slowly thereby causing the facepiece to collapse slightly. Upon the collapse of the facepiece, the wearer stops inhaling for at least 10 seconds while observing the facepiece. If the facepiece remains collapsed and no leakage is evident, it is likely that the respirator fit is good.

The positive pressure test is the converse of the above test. In this test the exhalation valve is closed off with a palm of a hand and the wearer gently exhales. An increase in pressure within the face should be felt.

The third test, which is applicable only for air supplied respirators and organic vapor cartridge or canister mask, involves the use of isoamyl acetate, or "banana oil", since it has a very pleasant, easily detectable odor resembling the smell of bananas. An isoamyl acetate atmosphere is created by saturating a piece of cotton or cloth with the liquid and passing it close to and around the respirator, near the sealing surface. Alternative procedures include using a stencil brush filled with the chemical or a commercially available ampule. A good fit is indicated by the wearer's inability to detect the banana oil.

Respirator testing should be performed upon issue of the device to each individual and periodically thereafter to ensure continued proper fit and performance of the respirator.

Inspection and Maintenance

As with any piece of equipment, inspection and maintenance are essential to respirator performance. Depending on the type of respirators worn, the work place, the work conditions, and the hazards present, a tailored maintenance program for respirator equipment should include, at the very least, cleaning and sanitizing equipment, regularly inspecting equipment for defects, and properly storing the devices.

The most important part of a maintenance program is the inspection. All respirators should be inspected before and after each use. If a certain device is not used routinely (i.e., escape respirators), then this piece of equipment should be inspected before and after each use and at least monthly by a safety officer. Inspection before use should include: 1) the connections; 2) the facepiece shape--it should not be bent or misformed in any way; 3) the headstraps, which should not be stretched out or loose; 4) the inhalation valve; and 5) the air-purifying elements' cartridges for puncture or blockage; similarly, air supply system hose lines should not be punctured or blocked. In addition, the regulator and any warning devices on self-contained breathing apparatus should also be examined to ensure that they function properly. Inspection carried out after each use (during cleaning) is especially important. Disassembly for cleaning provides a good opportunity to inspect each component of the device for flaws. Because respirators that are taken apart might be reassembled improperly, an inspection should be performed after the cleaning process.

When to clean and sanitize respiratory protective equipment depends largely on how often you use the equipment. If devices are used routinely, they should be cleaned, sanitized and inspected daily; on the other hand, if respirators are only used occasionally, the time frame for this maintenance may be weekly or monthly. At large waste sites respirator cleaning may be performed in a centralized cleaning and maintenance area using specialized equipment and trained personnel. The procedure for cleaning involves detergent or cleaner, a sanitizer, clean water and air drying. Cleaning may be performed by hand or done in a commercial dishwasher.

The last component of a successful maintenance program is storage. Respirators should be stored to protect them against dust, sunlight, extreme temperatures (hot and cold), excessive moisture, dangerous chemicals, and mechanical damage. Sunlight and extreme temperature can cause the rubber of the facepiece to crack and peel, which may result in a facepiece not fitting properly. Dust and excessive moisture may collect and accumulate in chemical cartridges causing damage and reducing the filtering capacity of the cartridge. Chemical contamination of a facepiece during storage could expose the wearer to an unnecessary hazard. Respirators stored on a workbench may be damaged by tools and heavy equipment. In order to avoid any of these storage problems, each cleaned respirator should be placed in a closed plastic bag and stored in a clean, dry area away from direct sunlight and heat. It is very important to store the respirators in a single layer and not in a position that would allow the rubber or plastic to become permanently distorted [11].

CHEMICAL PROTECTIVE CLOTHING

Hazard characterization has been addressed in other chapters of this book. In the preceding sections of this chapter, a wide variety of clothing and equipment was described as a means for minimizing worker exposure to hazards associated with waste sites. Barrier effectiveness should be a key concern in selecting chemical protective clothing (CPC).

Chemical hazards may be present in the form of solids, liquids and vapors. Each represents one or more challenges to PPE. Particulates, for example, are of special concern where solids are involved. Respirators, of course, provide protection from the inhalation of particulates. A less obvious need is to limit the transport of hazardous particulates from the workplace due to its accumulation in street clothing. Such particulates represent a hazard to those who associate with the worker and are not equipped with respirators. This is exemplified by the relatively higher incidence of asbestosis among the families of asbestos workers [12]. Similar concerns apply to lead and arsenic dusts, pesticide powders, and radioactive particulates. The current trend for countering such challenges is the use of porous, disposable garments which minimize heat stress problems while providing a barrier to the particulates.

Liquid and vapor challenges require a different level protection than that for the solids. Clothing containing a continuous (i.e., non-porous) layer of a plastic or rubber is necessary to isolate the worker from these hazards. Porous fabrics are unacceptable, although some promise is held for the new group of microporous, multilayer, breathable fabrics as splash protection.

Useful forms of the plastic or rubber (i.e., polymeric) materials for protective clothing are thin, flexible films, sheets, laminates and coatings. Many of these materials are inexpensive and in many cases can be reused multiple times. Furthermore a wide variety of candidate polymeric materials are available from which one or a select few may be applied to specific requirements. Items ranging from gloves to full-body ensembles can be made virtually air-tight and waterproof. Because of this, such items are often referred to as "impervious" clothing; however, with regard to chemical challenges, as opposed to air and water, such items may be far from impervious. Chemicals and chemical mixtures can absorb into and permeate clothing fabricated from polymers [13,14,15]. In the extreme, the chemicals may actually dissolve the clothing. Because of this, the chemical resistance of the clothing material to the chemicals of concern is a critical issue in the clothing/equipment selection process. This has been reviewed extensively elsewhere [6] and is the subject of the remainder of this chapter.

Permeation Theory

Permeation of a chemical through a polymeric material is a three-step process involving (1) the sorption of molecules of the chemical at the surface of the clothing exposed to the chemical, (2) the diffusion of the chemical through the material, and (3) the desorption of the molecules from the opposite or inside surface of the material. In this discussion, steps 1 and 3 will be considered fast relative to step 2; consequently, the diffusion step controls the rate of the permeation process. Classical permeation theory (Fick's laws) states that the permeation rate (mass/time/area) is proportional to the concentration gradient of the chemical across the material. The proportionality becomes an equation by the introduction of the diffusion coefficient. Thus,

(1) $J = - D\dfrac{dc}{dx}$

where J = mass flux, g, g/min/cm^2
    D = diffusion coefficient, cm^2/sec.
    c = the chemical concentration in the material, g/cm^3
    x = distance from the contacted surface, cm

The minus sign in Eq. (1) accounts for a decreasing c as x increases.

The diffusion coefficient is an intrinsic property of the chemical/material pair. It is a function of temperature and in some cases is dependent on the chemical concentration within the material. With the knowledge of D, one can estimate permeation rates for a range of material thicknesses and concentration gradients. Thus, it is worthwhile to determine diffusion coefficients for permanent/barrier pairs.

Diffusion coefficients are readily determined from the results of permeation testing. Such tests are typically conducted using a two-chambered cell in which the material of interest forms the partition between the two chambers. The challenge chemical or chemical solution is charged to one chamber (i.e., the challenge or upstream chamber) and the other chamber (i.e., the collection or downstream chamber) is monitored for the presence and concentration of the chemical that permeates the material. (The test procedure is described in more detail later.) Initially no chemical is detected. At some time, however, the chemical becomes evident; this is called the breakthrough time. Thereafter, the chemical appears at an increasing rate until a so-called steady-state is reached. This process is graphically depicted in Figure 1.

Two diffusion coefficients can be estimated from the data: the steady-state D and the time-lag D. The steady-state D_s is calculated using an integrated form of Eq. (1) and the assumption that the chemical concentration in the collection medium is maintained at essentially zero.

198

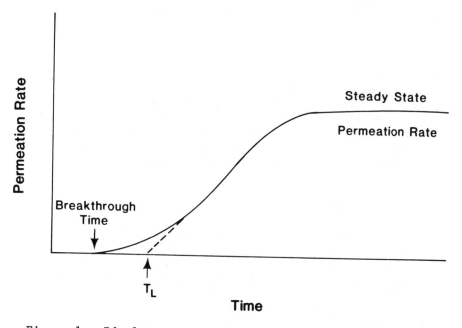

Figure 1.   Ideal permeation through a polymeric membrane.

Thus,

$$(2) \quad J = \frac{D_s\ C_2}{\ell}$$

or

$$(3) \quad D_s = \frac{J\ell}{C_2}$$

where C_2 is the saturation concentration of the challenge chemical in the material and ℓ is the material's thickness. C_2 is readily determined by immersion of a separate sample of the material until a constant weight is achieved. (Subtleties of immersion testing are described later.) The time-lag D_L is calculated according to eq. (4)

$$(4) \quad D_L = \frac{\ell^2}{6T_L}$$

where T_L is the time at the intercept of the extension of the steady-state line to the time axis [17].

Upon rearranging Eqs. (2) and (3), one notes that the steady-state permeation rate, J, is inversely proportional to thickness while the time-lag time, T_L, is proportional to the square of the thickness. Researchers have further shown empirically that breakthrough time is also proportional to the square of the thickness [18]. Thus, doubling the thickness of the polymer layer of an item of protective clothing will theoretically quadruple the breakthrough time. This finding has significant implications relative to the selection and specification of chemical protective clothing.

Ideal permeation was described above as a diffusion process in which the breakthrough time is followed by a period of smooth transition to a steady-state situation in which the permeation rate does not change with time. Ideal diffusion is likely to occur with most of the chemical/material pairs present on a hazardous waste site. It should be recognized, however, that deviations (i.e., anomalies) from the ideal may occur in a large fraction of the cases [13,16]. As the name implies, anomalous permeation is not predictable. There are, however, several general conditions under which the probability for non-ideal permeation is increased:

   * where there may be a reaction of the chemical with
     the plastic/elastomer of the CPC or some other
     component of the material. In some cases the
     reaction will lengthen the breakthrough time and
     reduce permeation rate by consuming chemical. In

200

other cases the reaction will reduce the barrier
effectiveness of the CPC by degrading its properties.
* where the chemical, merely by its being absorbed,
  changes the properties of the CPC. Many organic
  liquids are known to craze (produce surface cracks)
  in the hard, clear plastics used for lenses and face
  shields. Many of these same chemicals will soften or
  plasticize the clothing materials.
* where the chemical extracts components from the CPC
  materials. For example, leaching of plasticizer from
  PVC clothing will significantly affect its barrier as
  well as functional properties.

Nelson et al. [13], and Weeks et al. [19,20], discuss
this subject in more depth. Crank and Park [16,17] present
further discussion of polymers in general.
Permeation theory can provide significant insight to
clothing performance when data from testing are available.
It is often necessary to estimate CPC performance without
the benefit of test data. This may be especially true where
multi-component solutions are involved. At present there
are no established theories that provide a mechanism for
this activity although several investigators have approached
this problem from a basis of solubility parameter theory.
Furthermore, experience has led to the formulation of some
guiding principles relative to the probable chemical
resistance of clothing materials. The first is that, in
general, chemicals from the same family (e.g., the simple
alcohols, the primary amines, the alkanes, the aldehydes,
etc.) will tend to permeate a given CPC material at similar
rates and with similar breakthrough times. There are, of
course, exceptions. Other generalizations are listed below:

*Higher molecular weight members of a homologous series
 of chemicals permeate at slower rates than lower
 molecular weight members.
*Pendant groups (which increase the size of a
 molecule) tend to slow permeation relative to that of
 the simple molecule.
*Polar chemicals tend to permeate polar materials more
 rapidly than non-polar chemicals, and the converse is
 true.

Test Methods

The barrier effectiveness of a particular item of
clothing to a particular chemical/mixture is dependent on
the specific interactions between the clothing material and

201

the chemical/mixture. This in turn is determined by the formulation of the clothing material, its method of manufacture, and its thickness. Temperature and other conditions of use also influence clothing barrier properties. Finally, the composition of the chemical/mixture is of major importance since relatively small percentages of a second, third, etc., component can drastically alter the way in which a chemical interacts with a material. Thus, protective clothing selection decisions should be based on the results of testing of the chemical/clothing material pair whenever possible.

Immersion Test

Immersion of a clothing material in a chemical/solution followed by inspection for changes in appearance, strength, dimensions and weight is the easiest and perhaps most telling test of a clothing material. Weight change information is of particular interest since in general chemicals which are absorbed at levels of 10% or more are likely to rapidly permeate the material. There appears to be a correlation between weight change and breakthrough time [21]. Furthermore C_2 of Eq. (2) is directly estimated from the weight change of the material once it has reached a steady level.

ASTM Method D471-79 and ISO Method 2025 (International Organization for Standards) are standard methods for immersion testing. An important consideration when conducting such tests with multilayer clothing materials is that usually only the outside layer of the clothing should be exposed to the chemical/solution. The edges of such materials should not be exposed. Absorption of the chemical layers by sublayers or supporting fabrics that would not normally "see" the chemical would confuse interpretation of the results. The calculation of C_2 for a multilayer fabric may not be practical since it is likely to be difficult to determine the amount of chemical in each layer.

Finally, it should be noted that lack of change in an immersion test does not necessarily indicate that the material is a chemical barrier. This can only be determined by permeation testing as described in the following paragraph.

Permeation Test

Breakthrough time and permeation rate are determined by means of a permeation test. ASTM Method F739-81 was specifically developed for the evaluation of protective clothing materials [22]. The method uses a test cell which is divided into two chambers at the midline by the clothing material to be tested. The potentially hazardous chemical is placed in one chamber and the other chamber (*i.e., the collection chamber) is monitored for the chemical of interest. Of interest are the time the chemical is first detected (i.e., breakthrough time) and the subsequent rate of permeation. While the test is being conducted, the collecting medium must not interact with the clothing material; air, nitrogen, helium, or water are preferred collection media.

The detection of breakthrough is dependent on the sensitivity of the analytical method used for measuring the chemical in the collection medium. Typical, preferred analytical methods include gas, liquid and ion chromatography, analysis for total combustible organics, ultraviolet and infrared spectrophotometry and radioanalysis. The properties of the chemical, the sensitivity requirements for the test, and cost are the principal factors considered in selecting an analytical method. For relatively volatile chemicals, gas chromatography and infrared spectrophotometry are the preferred methods. Liquid chromatography is used for relatively nonvolatile organic compounds. Ion chromatography is particularly useful for inorganic acids and salts. Finally, radiolabeled compounds may be preferred where high sensitivity and specificity is required. Furthermore, if the compound of interest is readily available in radiolabeled form, radiochemical methods may be significantly less costly than the development and use of the other techniques.

Permeation testing of protective clothing materials has increased significantly during the past five years. The Journal of the American Industrial Hygiene Association has become the principal vehicle for dissemination of test findings. Also of note are the product catalogues of CPC vendors who have conducted such testing and use the results as the basis for their chemical compatibility charts.

Vision

Face shields and lenses, in addition to being chemical barriers, must provide clear, undistorted vision to the wearer. Hard, inflexible face shields and lenses fabricated from polymeric material may be subject to crazing (i.e., surface cracking) upon contact with certain chemicals. Crazing renders the surface foggy and can drastically reduce vision. Since chemical contact with the face shield or lens is more likely to occur in uncontrolled or emergency situations when reduced vision would be an additional severe hazard, shields and lens materials should be tested for resistance to chemical attack. Crazing can also reduce the impact strength of the material.

ANSI/ASTM Method F484-77 describes a procedure for determining stress crazing by chemicals. A second method for determining the effect of chemicals on clear plastics is by measuring the transparency of the plastic before and after exposure to the chemical; ASTM D1746 describes one such method.

Other Factors

Although the focus of this discussion is chemical resistance of clothing materials, the selection and use of protective clothing involves other factors of equal or greater importance. For example, gloves must provide the wearer some minimum level of dexterity, and fabrics must have some level of tear resistance. The relative importance of the performance factors is largely dependent on the work tasks to be carried out.

At present there is no standard, overall protocol for evaluating protective clothing or clothing materials for all the performance parameters of importance to workers on hazardous waste sites. Instead, individual tests appropriate for the evaluation of specific parameters must be selected from procedures promulgated by federal, military, and independent standards organizations. A compilation of pertinent materials is presented in Table IV. For completeness, the chemical resistance methods mentioned above are included in the table.

Table IV. Test Methods for Chemical Protective Clothing

Characteristic	Test
A. Chemical Resistance:	
1. Permeation Resistance...	ASTM F739-81: Resistance of Protective Clothing Materials to Permeation by Hazardous Liquid Chemicals
2. Swelling and Solubility.	ASTM D471-79: Rubber Property--Effect of Liquids
3. Strength Degradation....	ASTM D543: Resistance of Plastics to Chemical Reagents
4. Crazing.................	ASTM F484-77: Stress Crazing of Acrylic Plastics in Contact with Liquid or Semi-Liquid Compounds
5. Transparency............	ASTM 1746-70: Transparency of Plastic Sheeting
B. Strength:	
1. Tear Resistance/strength	ASTM D751-73: Testing of Coated Fabrics
	ASTM D412-75: Rubber Properties in Tension
	Fed. 191A-5102 (ASTM D1682): Strength and Elongation, Breaking of Woven Cloth: Cut Strip Method
	Fed. 191A-5134 (ASTM D2261): Tearing Strength of Woven Fabrics by the Tongue Method
2. Puncture Resistance.....	See reference 24
3. Abrasion Resistance.....	ASTM D1175: Abrasion Resistance of Textile Fabrics
C. Dexterity/Flexibility:	
1. Dexterity (gloves only).	See references 24 and 25
2. Flexibility.............	ASTM D1388: Stiffness of, Cantilever Test Method and
D. Aging Resistance:	
1. Ozone Resistance........	ASTM D3041-72: Coated Fabrics--Ozone Cracking in a Chamber ASTM D1149-64: Rubber Deterioration--Dynamic Ozone Cracking in a Chamber
2. UV Resistance...........	ASTM G27: Operating Xenon-Arc Type Apparatus for Light Exposure of Non-Metallic Materials--Method A--Continuous Exposure to Light

Literature Provided by Vendors

The most widely available sources of information on CPC are the product catalogues of the CPC manufacturers and vendors. These booklets contain descriptions of the types, sizes and varieties of CPC produced by each manufacturer. In most cases the basic materials of construction of the CPC are also included in the product descriptions. Many manufacturers also include information pertinent to the chemical resistance of their products or of the materials from which the products are fabricated. This information is generally in the form of tables of qualitative chemical resistance ratings or use recommendations for the products/materials and particular chemicals. A few vendors also provide information pertinent to abrasion, tear, etc., resistance but in general most catalogues do not address such application-related issues.

The chemical resistance ratings/recommendations are typically presented on a four-grade scale of "excellent," "good", "fair", and either "poor" or "not recommended". All vendors emphasize either in notes to their charts or in conversation that the ratings are provided to guide potential buyers in the selection of clothing. The charts do not nor do they claim to anticipate the applications to which the clothing will be put.

Other considerations when using such charts are:

*There is no standard test method on which the tables are based.
*The amounts of test data supporting the recommendations varies considerably from vendor to vendor. In many cases the vendor has no data and is relying on that provided by the supplier of the base polymer.
*The tables of many vendors are quite old. The ratings may not reflect changes in polymer formulation, raw materials suppliers, or fabrication processes that could effect chemical resistance.

Performance and Purchase Considerations

The performance of CPC as a barrier to chemicals is determined by the materials and quality of its construction. As discussed in the early sections of this chapter, the expected application is the key issued to address when selecting CPC. For example, a less durable

piece of clothing may be more than adequate for moderate
duration, mild activity (e.g., sampling), whereas it would
not endure more than five minutes of a vigorous, waste site
cleanup activity. Garment strength durability, and fit as
well as worker comfort must be addressed. Depending on the
application, chemical barrier effectiveness may be more or
less important than the physical attributes of the clothing.

When considering chemical resistance, three underlying
factors must be taken into account:

*In general, there is no such thing as "impermeable"
 plastic or rubber clothing.
*No one clothing material will be a barrier to all
 chemicals.
*For certain chemicals or combinations of chemicals
 there is no commercially available glove or clothing
 that will provide more than an hour's protection
 following contact.

Other considerations are:

*Stitched seams of clothing may be highly penetrable by
 chemicals if not overlayed with tape or sealed with a
 coating. Zippers, other closures, and interfaces
 (e.g., sleeve-to-glove) are also pathways for chemical
 ingress.
*Pinholes and areas where the polymer coverage is
 relatively thin can compromise barrier effectiveness.
*Although the generic names of the clothing material
 may be the same, there can be significant differences
 between the performance of the products of several
 vendors. This may be due to formulation or
 fabrication differences.

SELECTION OF PPE

When any piece or ensemble of PPE is to be utilized on
a hazardous waste site, the advantages and limitations of
the equipment should be carefully considered in regard to
the potential exposures to chemical and physical hazards.
Consequently, selection of the equipment should be performed
by an individual who is familiar with both the equipment and
the likely use conditions under which the PPE will be used.

Frequently, on hazardous waste sites the solutions and
mixtures of chemicals are unknown; often, the visible,
physical characteristic of the chemicals (solid, gas,
liquid) or the odor are the only available information.

Thus, it is difficult to assess the degree of hazard to which workers may be exposed. The state-of-the-art approach for selecting PPE in such cases is to initially assume the worst exposure condition and use the highest level of PPE. Then, as the chemical and physical agents on site are characterized, the PPE can be selected to match specific hazards. A good example of this approach is demonstrated by an Environmental Protection Agency advisory protocol for hazardous waste site entry, shown in Table V.

As illustrated on the left side of Table V, if any of the selection criteria listed under Level A was present on site, then Level A protective equipment must be used. As the concentration of contaminants, hazardous substances, potential for splash, and organic vapor levels are reduced, the level of protective equipment is lowered to Level C. It should be noted that Level D PPE is primarily a work uniform and should not be worn where there is potential for contamination to body parts through boots or when inhalation of gases or vapor is possible.

The selection of the most appropriate respirators has been discussed earlier. The selection of clothing requires decisions relative to the areas of the body which must be covered and the materials of construction of clothing. The extent of body coverage is a function of the hazard to be faced. Common practice is to use the minimum amount of PPE as necessary to provide protection to the worker. PPE can be burdensome and restrictive; minimizing PPE increases the likelihood that it will be worn and minimizes the loss in worker efficiency that typically accompanies PPE utilization. In addition, PPE can be expensive. Furthermore, each time PPE is used, it must be either disposed of or decontaminated and properly maintained; therefore, it is desirable to minimize time and costs directed to these activities. The determination of the amount of clothing which is appropriate for any given job is the purview of the industrial hygienist or safety engineers. These professionals must consider in their decision all aspects of the job to be done, the conditions under which it will be done, and the capabilities of the workers.

PPE USE

In order to obtain maximum performance from any item of PPE, it must be free of defects, in good operating condition and the wearer must understand the purpose of the item and how to use and care for it. PPE should be unpacked and

Table V. Environmental Protection Agency Site Entry Protocol

Level/Selection Criteria[1]	:	Eye/Face	:	Head

Level A
-Chemical concentration known--
  ABOVE SAFE LEVEL
-Extremely hazardous substance
(dioxin, cyanide)
-Skin destructive substance
-Confined spaces
-Organic vapor levels
  500-1,000 ppm
		See full body		Hard hat
				(under
				seat)

Level B
-IDLH and concentrations above PF
  provided by full mask, air
  purifying
- 19.5% O_2
-Skin contact unlikely to head and
  neck
-Organic vapor .5-500 ppm
		See respirator		Hard hat

Level C
-Known air concentration that PF
  will control in air-purifying
  mask
- IDLH
-No skin destruction
-Organic vapor 0-5 ppm
		See respirator		Hard hat

Level D -No measurable concentration -No exposure to splashes or inhalation		Safety glasses		Hard hat

[1]Meeting any of listed criteria requires that level of
protection.

Table V. Environmental Protection Agency Site Entry Protocol

Hands :	Feet :	Full Body	: Respirator
2 pairs gloves	Chem-resistant steel-toe, shank disposable booties	Fully-encapsulated chem-resistant suit w/disposable outer suit, gloves, boots	SCBA (Pressure-demand)
2 pairs gloves	Chem-resistant steel-toe, shank disposable booties	2-Piece suit w/hood or disposable suit	SCBA (pressure-demand)
2 pairs gloves	Steel-toe, w/shank disposable booties	2-Piece suit or disposable suit	Full-face air-purifying mask
1 pair gloves	Steel-toe, shank	Coveralls	None

inspected immediately upon its reception. This initial inspection is to check that the desired items were actually received and that the items are defect-free and operational. This inspection prevents the surprise of finding non-functional or inappropriate PPE in emergency situations or losing time while new PPE is ordered.

Following inspection, PPE should be stored in a cool, dry place with clear and definitive labels in order to prevent mix-ups that could result in the utilization of the wrong PPE for a given application. For example, gloves made from neoprene, butyl rubber, PVC, nitrile rubber can be similar in appearance, yet, there can be significant differences in the barrier performances of these materials.

At the time of use, each wearer should inspect the clothing prior to donning it. Again the objective is to identify tears, punctures, fabrication flaws or functional problems that could compromise the protection anticipated from the PPE. A post-donning inspection is essential for full-body encapsulating suits. This may be best carried out with the assistance of a second individual who is able to check closures and interconnections between, for example, gloves and sleeves, boots and pants, etc. Further reinspections should be performed throughout the work period, especially if the wearer has experienced significant contact with a chemical or suspects the integrity of the PPE has been breached.

Following completion of the work assignment or the work period, PPE is removed (doffed). A primary consideration in doffing is to avoid transfer of chemical that may be on the outside of the PPE to clean areas, skin, underclothing. It is common practice at waste sites to doff PPE at designated areas, in many cases following a preliminary decontamination of the PPE with soap and water. The EPA has developed comprehensive doffing procedures which address doffing, decontamination and disposal of contaminated PPE [23].

Decontamination and re-use of PPE is a matter of considerable interest and concern. At issue is any chemical that may have been absorbed by the PPE material. Is the chemical removed by the decontamination process? If not, what happens to this chemical during storage? Does the chemical continue to permeate the clothing such that the next time the PPE is donned, chemical is present on the inside surface? Researchers are only now beginning to address these problems; however, practitioners, must deal with the issue every day. Some have opted for the use of inexpensive, single-use disposable clothing whenever possible. Such clothing is not universally applicable, however, and the use of more expensive PPE may be required. Some full ensembles can cost $1,000 or more versus less than $10 for some disposables while the most expensive gloves are in the range of $20-25/pair versus less than a dollar a pair for the less expensive gloves. Obviously, there is an economic incentive to re-using the more expensive items; the challenge is to ensure that these items are effective and have no intrinsic hazard the second time they are worn.

SUMMARY

Since engineering controls are not readily implemented at hazardous waste sites, PPE combined with good work practices are the primary means for minimizing the exposure of workers to hazardous chemicals. PPE ranges from respirators, to supplied air systems, to gloves, to full-body encapsulating ensembles. Proper selection of PPE requires careful assessment of the risk hazard. This includes the chemicals involved, the skills of the workers, the tasks and the duration of potential exposures. PPE must then be selected on the basis of its demonstrated performance under such conditions. With regard to clothing, chemical resistance is a key concern and it must be recognized that there is no universal barrier material. Once PPE has been selected and purchased it should be inspected for construction flaws and function. Workers must be instructed as to the use and limitations of PPE. Re-use requires special attention to decontamination and storage.

REFERENCES

1. "Environmental News Superfund Status Report," Office of Public Affairs, Environmental Protection Agency, Washington, D.C., (January 10, 1984).

2. Streng, D.R., W.F. Martin, D. Weitzman, G. Kleiner, and J. Gift. "Hazardous Waste Sites and Hazardous Substance Emergencies--Worker Bulletin," U.S. Department of Health and Human Services, Publication No. 83-100 (1982).

3. Clayton, G.D., F.E. Clayton. Patty's Industrial Hygiene and Toxicology, 3rd Revised Ed., (vols. 1-3, Wiley-Interscience).

4. Mackison, F.W., R.S. Stricoff, L.J. Partridge. "NIOSH/OSHA Occupational Health Guidelines for Chemical Hazards," DHHS (NIOSH) Publication No. 81-123 (1981).

5. Accident Prevention Manual for Industrial Operations, 7th Ed. (National Safety Council, 1974).

6. Schwope, A.D., P.P. Costas, J.O. Jackson, and D.J. Weitzman. Guidelines for the Selection of Chemical Protective Clothing. (Am. Conf. of Govt. Ind. Hygienists, Cincinnati, OH, 1983.

7. Pritchard, John A. A Guide to Industrial Respiratory Protection, U.S. Energy Research and Development Administration (1977).

8. General Industry OSHA Safety and Health Standards (29 CFR 1910.134: Respiratory Protection, U.S. Department of Labor, Occupational Safety and Health Administration (OSHA 2206, June 1981).

9. J.B. Olishifski, P.E. McElroy and F.E. McElroy. Fundamentals of Industrial Hygiene, National Safety Council (1977).

10. NIOSH Certified Equipment List, U.S. Department of Heath and Human Services, DHHS Publication No. 83-122 (1983).

11. L.R. Birkner. Respiratory Protection Program--A Manual and Guideline, American Industrial Hygiene Association (1982).

12. Peters, G.A. and B.J. Peters. Sourcebook on Asbestos Diseases (Garland STPM Press, NY 1980), p. B-7.

13. Nelson, G.O., B. Lum, G. Carlson, C. Wong, and J. Johnson. "Glove Permeation by Organic Solvents," Am. Ind. Hyg. Assoc. J. 42(3): 217-225 (1981).

14. Sansone, E.B. and Y.B. Tewari. "The Permeability of Protective Clothing Materials to Benzene Vapor," Am. Ind. Hyg. Assoc. J. 41(3): 170-174 (1980).

15. Williams, J.R. "Chemical Permeation of Protective Clothing," Am. Ind. Hyg. Assoc. J., 41(12): 884-887 (1980).

16. Crank, J. and G. Park. <u>Diffusion in Polymers</u>, (Academic Press, NY 1968).

17. Crank, J. <u>Mathematics of Diffusion</u>, 2nd ed., (Claredon Press, Oxford, 1975).

18. Todd, W.F., A.D. Schwope, and G.C. Coletta. "Benzene Permeation Through Protective Clothing Materials," paper presented to the 72nd AICHE Annual Meeting, San Francisco, Nov. 1979.

19. Weeks, R.W., Jr., and M.J. McLeod. "Permeation of Protective Garment Materials by Liquid Benzene and by Tritiated Water," Am. Ind. Hyg. Assoc. J., 43(3): 201-211 (1982).

20. Weeks, R.W., Jr., and M.J. McLeod. "Permeation of Protective Garment Material by Liquid Halogenated Ethanes and a Polychlorinated Biphenyl," Los Alamos National Laboratory Report #LA-8572-MS, 1980.

21. Stampfer, J.F., M.J. McLeod, M.R. Betts, and S.P. Berardinelli. "The Permeation of Eleven Protective Garment Materials by Four Organic Solvents," submitted to Am. Ind. Hyg. Assoc. J. (1984).

22. Henry, H.W. and C.N. Schlatter. "The Development of a Standard Method for Evaluating Chemical Protective Clothing for Permeation by Liquids," Am. Ind. Hyg. Assoc. J., 42(3): 202-207 (1981).

23. See Appendix G of Reference 6.

24. Coletta, G.C., A.D. Schwope, I. Arons, J. King and A. Kivak. "Development of Performance Criteria for Protective Clothing Used Against Carcinogenic Liquids," Arthur D. Little, Inc., Report to NIOSH under contract 210-76-0130, October 1978.

25. Forsberg, K., A.L. Linnarson, K. Olsson, and L. Sperling. "Development of Safety Gloves. Gloves for Printers," ASF Contract 80/220, ERGOLAB Report S 81:10, Stockholm/Goteborg, Sweden, Nov. 1981, (translated from Swedish).

10

HEAT STRESS IN INDUSTRIAL PROTECTIVE ENCAPSULATING GARMENTS

Ralph F. Goldman, Ph.D.
Chief Scientist
Multi-Tech Corporation
One Strathmore Road
Natick, MA 01760

Heat stress represents an imbalance between the heat produced by an individual and the heat loss allowed to the environment. The latter is as frequently controlled by the clothing worn as by any combination of environmental conditions; there is no single temperature or combination of temperature and humidity at which heat stress can be said to begin. Heat stress has occurred in men working very hard in the snow, although it is usually not recognized as such but thought to be some mysterious ailment. Typical heavy, outdoor winter clothing ensembles, e.g., clothing insulation of 4 clo (page 226), may only allow heat loss of 2.5 kcal/hr per $^{\circ}$C difference between the skin of the wearer and the ambient environment. Even at -40°, the maximum heat exchange by radiation and convection (H_{R+C}) through such clothing would be less than 200 kcal/hr; i.e., 2.5 kcal/hr. $^{\circ}$C times the 75°C difference between a warm 35°C skin temperature and the ambient of -40°. While some additional heat would be lost by respiration, since heat production for sustainable high activity is about 500 Watts (425 kcal/hr) about half of the total heat production would require sweat evaporative cooling (E_{req}) for it to be eliminated, but such clothing stringently limits the wearer's maximum evaporative cooling (E_{max}).

Using an average value of 58 kcal increase in body heat storage for each 1°C increase in mean body temperature [t_b = 1/3 skin temperature (t_{sk}) + 2/3 deep-body temperature (t_{re})] for a 70 kg man (i.e., specific heat of body = 0.83 kcal/kg. $^{\circ}$C), it seems clear that accumulation of heat storage (ΔS) which would lead to body temperatures above 39°C could occur within a few hours, even at -40.

The question of human heat balance can be analyzed by a well-defined heat balance equation. A major factor that must be considered is the metabolic energy production (M); the heat production required for a given task is a function of the total weight (body plus any load) moved, the efficiency with which this total weight is moved, and the rate of movement. The nature of the terrain over which the weight is moved is also involved. Walking on soft sand may double the energy cost for a given speed and weight compared to walking on a smooth surface. Climbing stairs, or any lift (grade) work, is also very demanding in energy cost increases.

Another major factor, the non-evaporative heat exchange (H_{R+C}), is a linear function of the difference between the wearer's skin temperature (t_{sk}) and the ambient air temperature (t_a). Similarly, the maximum possible sweat evaporative cooling (E_{max}) is a linear function of the difference between the vapor pressure of sweat on the skin (P_s) and the ambient vapor pressure ($\phi_a P_a$); the latter is the relative humidity (ϕ_a) times the saturated vapor pressure of air (P_a) at ambient temperature (t_a). The heat exchange with the environment depends not only on the skin and ambient environmental conditions but also on the clothing, and the extent to which it limits the heat exchange between the skin and the ambient environment. Chemical protective clothing tends to be quite limiting both because of its insulation (clo) and its reduced moisture vapor permeability (i_m). Air motion (WV) also plays a key role in controlling these non-evaporative and evaporative exchanges, primarily by the extent to which it reduces the insulating still air layer (I_a) at the interface between the clothing surface and the ambient environment, although the intrinsic insulation of the clothing per se (I_{clo}) can also be reduced by wind penetration. A fourth environmental factor (in addition to t_a, $\phi_a P_a$, and WV) which frequently must be considered is the radiant heat load (H_R) produced by the sun, by such high temperature (shortwave, infrared) heat sources as blast furnaces or arc lights, or by such lower temperature "black body" sources as radiators, warm pipes, walls or ceilings. The mean radiant temperature (MRT) is an integrated expression of the average radiant temperature.

THE HEAT BALANCE EQUATION

The factors introduced so far, form the key elements in the following heat balance equation for the human body:

216

<u>Eq. 1</u> $(M - W_{ex}) + (H_{R+C}) - E_{req} - \Delta S = 0$

where: M   = energy production (measured by oxygen
                consumption)
        W_{ex} = external work
        H_{R+C}= the net exchanges by radiation and convection
                between the body and the environment
        E_{req}= the required evaporative heat loss
                established by $[(M-W_{ex}) + (H_{R+C})]$
        ΔS = any change in body heat content.

Additional terms are sometimes included such as the heat
exchanges by respiration, involving both humidification and
heating of the inspired air, and diffusional evaporative
heat losses from the skin; these respiratory and diffusional
losses generally amount to about 25% of metabolic heat
production at rest, and are most often ignored during work
in the heat.
     The amount of evaporative cooling (E_{req}) in the heat
balance equation would ideally be much less than the maximum
evaporative cooling that can be obtained through the
clothing (E_{max}); i.e.: $E_{req} = [(M-W_{ex}) + (H_{R+C})]$
$<< E_{max}$.
As long as the heat losses are less than $(M-W_{ex})$, and all
the evaporative cooling required can be obtained, no change
in body heat storage (ΔS) is required. If heat loss by
radiation and convection is greater than $(M-W_{ex})$ a heat
debt will be incurred [34]. Changes of ± 25 kcal in body
heat content are probably not detectable by an individual,
but an accumulation of body heat storage approaching 60-80
kcal generally results in the individual being unwilling to
continue. Thus, satisfaction of the heat balance equation
with minimal heat storage is a necessary condition for
comfort and continued work. It is not, however, a
sufficient condition; discomfort in the heat is largely
generated by the sense of skin wettness or dampness. The
sensation of wettness at the skin can be directly
calculated as the ratio E_{req}/E_{max} [16]; i.e., the body
will produce enough sweat to meet its evaporative
requirements so that the relative humidity at the skin
(alternatively defined as the "percent skin wettedness") can
be calculated simply as the ratio E_{req}/E_{max}. If E_{req}
is 50 Watts (i.e., $M-W_{ex} + H_{R+C} = 50$), and $E_{max} = 100$
Watts, then the skin need be only 50% sweat-wetted, or skin
relative humidity will be 50%; expressed more precisely, the
average skin vapor pressure (P_s) will be 50% of the
saturated vapor pressure of water (= sweat) at t_{sk}.
     As the maximum evaporative cooling (E_{max}) approaches
or is less than the required evaporative cooling (E_{req}),
the body cannot obtain the required evaporative cooling.
The maximum evaporative cooling may be limited by the

clothing, or by a high ambient vapor pressure, or even by very low ambient air motion. A necessary condition for comfort is that the relative humidity at the skin (i.e., % sweat wettedness) be less than about 20%. Note that, as discussed later, this ratio has been used as a heat stress index [HSI; [6]]. Increasing levels of sweat wettedness are associated with increasing heat discomfort. One must carefully screen and select workers for physical fitness at conditions requiring 60% sweat wettedness; at about that level, sweat will begin dripping off the skin. In general, a 60% sweat wettedness (i.e., skin relative humidity of 60% as defined by the ratio E_{req}/E_{max}) will be about the highest acceptable level for even a well-motivated, very fit and well-acclimatized civilian workforce.

So far, the heat balance equation has been presented and its importance identified in determining whether or not a given combination of factors results in heat stress. While the reader may be concerned at the rather casual treatment of respiratory and diffusional heat losses, it must be recognized that the heat balance equation itself represents an approximation. Considering the variability in individual size, in physical and physiological states between workers, in the clothing worn and in its fit on a given individual, it is obvious that any human heat balance equation is inherently not a precision statement. Thus, this approach clearly falls within a GEGU--good enough for general use--category. This GEGU acronym will be used to emphasize adequacy, albeit imprecision, at a number of points in the subsequent discussion.

The six key parameters involved in the calculation of the heat balance equation have now been identified. Four are environmental factors: the air temperature (t_a), the ambient air motion (WV), the ambient vapor pressure ($\phi_a P_a$), and the mean radiant heat temperature (MRT). The other two factors, which are subject to behavioral temperature regulation, are the task workload (and the associated heat production of the worker) and the clothing worn by the worker. Each of these factors will be addressed in turn.

THE SIX KEY FACTORS

1.   Ambient Air Temperature (t_a)

Surrounding every physical object is a surface film of trapped air. This "surface still air layer" contributes a very significant part of the insulation surrounding a human body (usually over 50% when dressed for indoors). This insulating air film is altered by air movement and this accounts for the perceived difference between the ambient air temperature as sensed by a still hand, and the ambient air temperature sensed when the hand is in motion. If one simply hangs a thermometer in ambient air, any radiant heat in the environment will be absorbed by the thermometer and effectively trapped there by the surface film layer around the thermometer. Accordingly, in order to measure the true psychrometric properties of air the thermometer must be ventilated, to minimize any surface still air film. This is done either by having the thermometer swung on the end of a chain (a sling psychrometer) or by having air pulled across it by a fan (an aspirated psychrometer). If the work site is close to an intense radiant heat source (smelter, glass furnace, etc.) special shielding for the thermometer bulb, or a specially shielded sensor, may also be required.

2.   Ambient Vapor Pressure ($\phi_a P_a$)

The ambient vapor pressure is usually determined from the measurement of a "wet bulb" thermometer temperature. Most psychrometers, sling or ventilating, simply pair two identical thermometers, with one bulb mounted a few centimeters below the other, and equip the lower bulb with a wettable cotton wick; hence the terminology dry bulb ($t_{db} = t_a$) and wet bulb (t_{wb}) temperature. The wick is saturated with water prior to aspirating or slinging the psychrometer. Because of both the omnipresent potential for convective heat gain from the air by the evaporatively cooled wet bulb thermometer and the potential radiant heat regain, ventilating the wet bulb thermometer to the appropriate air motion [air movement $\geq$4.5m/s (900 fpm) past the wet bulb wick] is of critical importance.

Some environmental physiologists have argued, correctly, that use of a psychrometric wet bulb to represent the potential evaporative cooling available to a worker is unreasonable--unless the worker is somehow to be ventilated or slung by the heels to produce a high air velocity across his 100% sweat-wetted skin. Accordingly, some of the environmental indices, to be discussed subsequently, incorporate a "natural"--or non-psychrometric--wet bulb (t_{nwb}).

Figure 1 presents a standard temperature-vapor pressure (Moliere) diagram for air at sea level, which can be used to

219

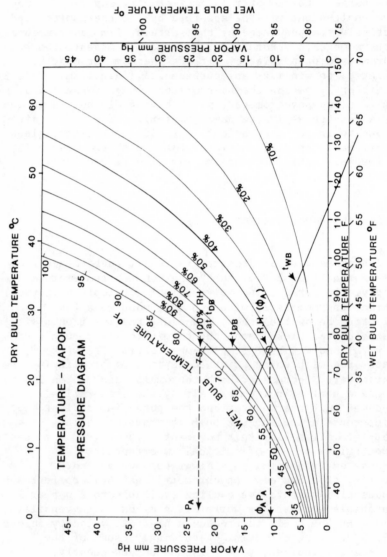

Figure 1. Psychometric chart for air at sea level showing relationships.

220

convert the measured dry bulb and wet bulb temperatures to a
relative humidity; in turn, this can be converted to the
ambient vapor pressure which is the key environmental
parameter required for calculation of the maximum
evaporative cooling capacity. Alternatively, tables of wet
bulb depression at a given dry bulb temperature, or a
"psychrometric slide rule", can be used to obtain the
percent of relative humidity. Note that, because
evaporation is a function of the difference between the skin
and the ambient vapor pressures, evaporation of sweat from
the skin can occur at 100% relative humidity as long as the
air temperature is less than the temperature of the skin;
thus ambient relative humidity <u>per se</u> is of little interest
for heat balance.

The point of intersection of the wet bulb and dry bulb
temperature lines on the psychrometric chart (cf. Fig. 1)
identifies the relative humidity; the uppermost curve on the
graph represents the 100% relative humidity line. At 100%
relative humidity, the ambient air is saturated (i.e.,
cannot take up any more moisture), so the wet bulb and dry
bulb temperatures will, of course, be equal; no evaporation
can occur unless air temperature increases. Observation of
the psychrometric chart also indicates that the vapor
pressure can be determined from this point of intersection
of the dry bulb and wet bulb lines; it is simply read from
the Y axis in either kilopascals (the SI unit) or, the more
familiar, mmHg unit. In the example drawn on the
psychrometric chart (Figure 1) the ambient vapor pressure is
one kilopascal or 7.5 mmHg. Note that at the indicated dry
bulb temperature of 25.5°C, the ambient vapor pressure at
100% relative humidity would be about 25 mmHg; i.e., P_a,
the saturated vapor pressure would be 25 mmHg. Multiplying
this P_a by the 30 percent relative humidity indicated on
Figure 1 (by the intercept of the dry bulb and wet bulb
lines, ϕ_a) yields 7.5 mmHg, the ambient vapor pressure;
simply stated, the air is holding 30% of the total moisture
that it could hold when saturated at t_{db}.

Under conditions where there is little or no
requirement for sweat evaporative cooling (i.e., low work
with light clothing at comfortably cool to colder
temperatures), ambient vapor pressure is of little concern
since there is no requirement for evaporative cooling.

3.   Air Motion (WV)

The importance of the surface still air film as
insulation has already been noted. For a cylinder the size

of the human body, with low air motion the still-air layer
film at the surface becomes a very significant contributor
to the total insulation of a human. Indeed, wearing a
long-sleeved shirt and trousers under still-air conditions,
the contribution of the external air film (I_a; = 0.8 clo)
equals or exceeds the intrinsic insulation (I_{cl}; = 0.6
clo) of the shirt and trousers. Air movement then is a
major factor in heat transfer from the body to the ambient
environment, with or without clothing. Even with an
impermeable, encapsulating chemical protective clothing
system, the external surface-air film (I_a) is still
important unless reduced by wind.

Measurement of air movement requires sophisticated
instrumentation, or sufficiently high but non-turbulent air
motion that simpler field devices can be used. Thus, the
usual approach for determining air motion indoors and
outdoors, is to estimate it. One frequently sees air
movement specified as 50 feet per minute (fpm), or as a
seemingly more exact 44 fpm which is simply one-half mile
per hour. Such values are simply GEGU estimates. With
ambient air motion at about 0.13m/s (25 fpm), the natural
convective air motion which results from the temperature
difference between skin and air temperature becomes the
primary factor; thus, lower ambient air motion is
meaningless. The insulation of the surface still air layer
at that low air motion is about 0.8 clo units for a human
body.

In addition, if the worker is moving, body motion
generates an "effective" air velocity [10]. Newburgh
suggested that the effective wind velocity (V_e) generated
during activity could be estimated from the heat production
of the worker, using the MET unit of metabolism (one MET
equals an average, "resting" heat production of 50
kcal/m^2hr) he suggested the relationship:

Eq. 2 $V_e = .07 (MET-0.85)$

where: V_e = effective wind motion generated (in m/s)
          and 0.85 = a "sedentary" MET level

Fanger suggested that the effective air velocity could be
calculated as $0.1 + 0.4 (MET-1)^{0.5}$, which gave a heat
transfer coefficient (h_c) for average indoor clothing of
$12.1 (V_e)^{0.5}$ (V_e in m/s; hc in watts/m^2°C). Such
refinements are unnecessary for most practical work but
should be kept in mind under conditions of very low ambient
air movement. The effective air velocity is more important
with light clothing and is particularly important with air
permeable, chemical protective clothing as will be addressed
further in the section on clothing.

4. Mean Radiant Temperature (MRT)

Indoors, the temperature of the wall, windows, floor and ceiling are usually considered equivalent to the temperature of the air; however, thermal radiation can be a major contributor to discomfort and heat stress when an individual works near a large window or on the top floor of a building with an uninsulated roof. Thermal radiation is a major concern in industrial settings with such large, high-temperature heat sources as ovens, arc furnaces, etc.

The mean radiant temperature of an environment is an integrated value, representing the uniform surface temperature of a radiantly black enclosure in which an individual would exchange the same amount of radiant heat as he does in the actual, non-uniform, radiant environment. The usual mean radiant temperature measuring device is a hollow, thin-shelled, 15 cm (6-inch) copper sphere, painted flat black; the globe temperature (t_g) is measured at the center of the inside of the sphere. The mean radiant temperature (MRT) is calculated from the globe temperature [3], as a function of the ambient wind velocity (WV), by the equation:

Eq. 3

$$MRT = t_g + k \ (WV)^{0.5} \ (t_g - t_a)$$

where: $k = 2.2$ for $t(°C)$ and WV (m/s)
or $k = 0.157$ for $t(°F)$ and WV (fpm)

5. Metabolic Heat Production

As indicated above, the metabolic heat production of an individual is frequently expressed as so many Watts (= 1.163 kcal/hr) per square meter of surface area. A typical "standard" male will weigh 70 kg (154 lb), stand 174 cm (5 foot 8-1/2") tall, and therefore have 1.8 square meters (19.5 ft^2) of body surface area. A "standard" female will weigh about 57 kg (125 lb), stand 164 cm (5 foot 4-1/2") tall and have a body surface area of 1.6m^2. The resting heat production of a standard adult male will be about 105 watts, which can be calculated quite nicely using either 1.5 watts/kg of body weight (1.3 kcal/hr kg) times 70kg, or 58 Watts/m^2 (=1 MET=50kcal/hr m^2) times 1.8m^2. Because of the larger amount of body fat (which is relatively inactive) in females, their resting heat production (about 85 Watts, or 80% that of the standard male) cannot be approximated by using the standard, surface area based MET unit; it can,

223

however, be approximated using the 1.5 Watts/kg of body
weight relationship. Since the heat production requirement
for almost all physical work is a function of the weight
moved (e.g., kg of body weight) and the female must move the
weight of her fat, the weight-based equation presented below
for estimation of energy cost (i.e., heat production)
applies equally well for men and women.

There is little need to initiate new energy cost
measurements; extensive tabulations exist for the heat
production associated with almost all forms of human
activity [50]. Alternatively, if the primary element
involved in the activity is walking, the following equation
has been developed [21] (and validated across a wide range
of studies) to predict the heat production:

Eq. 4 $M = 1.5(W) + 2(W+L)(L/W)^2 + (\mu)(W+L)(1.5V^2$
           $+ 0.35\ VG)$

where: W = body weight (kg); L = load carried (kg);
        V = walking velocity (m/s); G = grade (%); and
        μ = a non-dimensional "terrain coefficient" ranging
        between a value of one, for a hard surfaced floor,
        to a value of two for soft sand.

The first term in the above equation simply expresses the
resting energy cost of 1.5 Watts/kg of body weight, which
works GEGU for males and females. The second term simply
represents the energy cost of standing with a load on the
back. If some portion of the load carried is not borne on
the torso, adjustment should be made for the inefficiency of
loading; each pound carried by hand costs roughly the
equivalent of two pounds on the back, and each pound of
footwear worn while walking is equivalent to five pounds
carried on the back [56]. The importance of keeping the
weight of protective footwear at a minimum can be seen from
this relationship.; e.g., chemical protective boot covers
which weight two pounds have the equivalent effect of an
added ten pounds of backpack weight during walking. The
weight of headwear should, in theory, be incremented by
about 30% to equate it to a back-carried load, but generally
this is too small an adjustment to require consideration.
However, it should be recognized that the frequent
complaints of the weight of protective headwear are apt to
stem from too high a center of gravity, and the accompanying
torque and momentum changes with body motion which result in
a higher perceived weight, rather than from the actual
weight of protective headwear per se.

The next term in the equation indicates that energy
cost goes up as a function of the square of the walking
speed and introduces a coefficient (μ) to adjust for the
nature of the terrain being traversed [33]. Any smooth,

224

hard surface, be it a treadmill, floor or blacktop road
requires essentially the same energy cost and   for these is
assigned a multiplier of 1. Terrain coefficients (i.e.,
relative multipliers) to adjust for the energy costs of
walking at a given speed across other terrain surfaces are:
1.1 for a gravel road; 1.2 for light brush; 1.3 for packed
snow or ice; 1.5 for heavy brush; 1.8 for swampy terrain;
and 2.1 for soft sand. A task requiring heat production
below 5 kcal/min would be considered average work, and would
require a heart rate between 75 and 100 beats/min. to
deliver sufficient oxygen to the working muscles. Moderate
to hard work tasks would require up to 7.5 kcal/min (or 523
Watts) of metabolic energy cost; these would require heart
rates in the range of 100-125 b/min. Heavy work,
sustainable for about one hour by an individual of average
fitness, would correspond to an energy cost of about 10
kcal/min. (700 Watts) and require heart rates between 120 to
140 b/min. Finally, physical work which corresponds to a
physiological energy cost of about 15 kcal/min. (or 1,050
Watts), would present a work level that might be sustained
by an average, young individual for only about 10 minutes,
and would push heart rates up to 160 to 180 b/min. If the
same physical work were done under hot conditions, the
metabolic energy costs would be slightly, if at all,
increased initially, but the heart rates would increase
rapidly and dramatically as the burden of transferring heat
from the working muscles to the skin increased (see page
259, "Convergence"). Eventually the heat production would
also increase dramatically, as the worker became
increasingly uncoordinated, dizzy and, eventually, unsteady
and staggering if work continued.

Note that these work rate-time limitations refer to
reasonably fit young adult males; they really represent
given percentages of such individuals' maximum oxygen uptake
($\dot{V}O_{2max}$), or "maximum work capacity". Industrial tasks
seldom demand more than 5 kcal/min, or roughly one-third of
an average young adult male's ($\dot{V}O_{2max}$). A number of
studies have shown that the voluntary hard work level
adopted by individuals who must sustain such work for at
least three to five hours corresponds to about 45% of their
capacities [~7 kcal/min for fit young men [36]]. Although
sustained, high physical work demands are rare in industrial
work, individuals can sustain work demanding 60% of their
individual $\dot{V}O_{2max}$ capacity for about one hour, 75% of
their capacity for about 30 minutes, 85% of their capacity
for about 15 minutes and working at their maximum oxygen
uptake will be exhausted in about six minutes, almost by
definition of $\dot{V}O_{2max}$.

6. Clothing

The unit used to express clothing insulation (18) is of relatively recent origin; it was first proposed in 1941 by Dr. A. P. Gagge that the insulation of a typical business man's wool suit of the late 1930's be taken as one clo value of insulation. The mathematical value assigned to one clo was derived by calculating the potential difference for non-evaporative heat transfer from the human body to the ambient environment, and dividing it by the desired heat flow to calculate the resistance of the clothing worn which, it was assumed, allowed heat balance to be established for the wearer. This desired heat flow was assumed to be the resting heat production (M), one MET (58 Watt/m^2 or 50kcal/m^2hr), less the 25% of the resting heat production lost from the body by respiration and by evaporation of body moisture, diffusing through the semipermeable skin and evaporating to the air. Using 33°C as a comfortable skin temperature and 21°C as a standard room temperature (in the early 1940's) produced a 12°C driving force for non-evaporative heat transfer. Dividing by the desired heat flow of 38kcal/m^2hr (i.e., 75%M) provided a total conductance for the clothing plus the external air film at the clothing surface of 0.32°Cm^2hr/kcal. In subsequent studies on nude men in still air, the still-air surface film conductance was evaluated at 0.14°Cm^2hr/kcal, which left the intrinsic conductance of the heavy business suit of the early 1940's as 0.18°Cm^2hr/kcal. Taking the reciprocal of this conductance established the value for one clo unit of insulation (I) as 5.55 kcal/m^2hr°C (or 6.45 watt/m^2°C). Under these conditions, the air temperature and mean radiant temperature were identical and the clo value is, in fact, the combined insulation against radiative and convective heat transfer. For conditions where air and mean radiant temperature are not very closely equal, one can simply substitute the adjusted dry bulb temperature (t_{adjb} = (MRT + t_a)/2). For ease in calculation, one may use the insulation per man rather than per m^2; then, for the average adult male who has a surface area of 1.8m^2, one clo of insulation results in a heat loss of 10 kcal/hr°C (11.63 Watts/°C), two clo of insulation requires the transfer of 5.8 Watts (5 kcal/hr) per °C difference between skin and air temperature, etc. The insulation of a material is almost always a linear function of its thickness, with 1.57 clo of insulation provided by each cm of material thickness (4 clo per inch); in essence, a 6.5 mm (1/4") thick blanket provides one clo unit of intrinsic insulation (I_{cl}) (45). This same general approach applies

to the insulation used in building construction; the clo
unit of clothing insulation is equal to 1.14 of the R units
used in building insulation.

Thus, in calculating non-evaporative heat transfer, one
simply needs to know the total clo value of the clothing
worn and the surface still air film (I_a) trapped at its
surface. The total insulation value of an ensemble is
measured with a life-sized, heated (and when desired,
sweating by means of wetted cotton "skin") manikin whose
heating wires are distributed throughout the skin to produce
a human skin temperature pattern. Such manikins also have
temperature sensors distributed throughout their surface to
measure an average skin temperature and a thermostat to
demand sufficient heat to maintain a constant average skin
temperature. When such manikins are run in a controlled
temperature environment in steady state, the amount of heat
demanded to maintain a constant skin temperature is exactly
equal to the amount of heat lost. This allows direct
measurement of the total insulation value of any clothing
ensemble so tested, using the 6.45 W/m^2°C (5.55
$kcal/m^2hr$°C) defining value of one clo of insulation [57].

For cold weather conditions, one simply needs to know
the heat production of the individual and the clo value of
his insulation, in order to calculate whether the heat loss
from the body will match the 75% of heat production
available for non-evaporative losses. If it does not, any
excess heat loss demand will be withdrawn from the body, in
which case body cooling results; if less heat is lost than
is produced, heat storage by the body must ensue unless the
body can lose heat by evaporative cooling.

When less heat is lost through the clothing insulation
than required to match the heat production at rest or work,
then the 25% of resting metabolic heat production lost by
respiration and diffusion of moisture through the skin, and
its evaporation, must be supplemented with actual sweat
evaporation; i.e., the normal 6% relative humidity of the
skin (or 6% diffusion "sweat-wetted area", equivalently)
must be increased by production of sweat by the body. Note
that if sweat cannot be evaporated, no cooling benefit is
derived; the individual simply dehydrates at a more rapid
rate unless adequate drinking water is taken. This is a
frequent problem with chemical protective clothing; e.g., a
well heat-acclimatized individual will produce more sweat
then one who is not acclimatized [19] but, if the clothing
worn is a barrier to sweat evaporation, heat acclimatization
is of little benefit and may simply contribute to more rapid
dehydration and earlier onset of heat exhaustion [27].

Table I gives representative values for the type of
chemical protective clothing ensembles that might be worn by
industrial workers. Note that these values are for normally

Clothing Assembly[1]	Insulation[2] (clo)	Permeability (i_m)	Index Ratio (i_m/clo)
Category I-- **Everyday Clothing**			
Long-Sleeved Shirt + Trousers	1.41	0.37	0.26
Supplemented with:			
a. Safety helmet	1.49	0.37	0.25
b. Safety gloves	1.48	0.36	0.24
c. Mask, hood	1.56	0.29	0.18
d. Air back pack	1.45	0.34	0.23
e. Plastic apron	1.50	0.28	0.18
f. a+b+c+d+e	1.70	0.24	0.14
Category II-- **Charcoal-in-foam[5]**			
a. Worn alone, open[3]	1.65	0.40	0.24
b. Worn alone, closed[4]	1.92	0.32	0.18
c. Worn over long shirt and trousers, open	1.97	0.42	0.21
c. Worn over long shirt and trousers, closed	2.30	0.35	0.15
Category III-- **Impermeable (butyl)**			
a. Worn alone, open	1.58	0.12	0.08
b. Worn alone, closed	2.05	0.09	0.04
c. Worn alone, w/wetted terry coverall	2.05	0.27	0.13

[1]Includes underwear (T-shirt, shorts), socks and shoes; values estimated from comparable military assemblies.
[2]All values given at 0.3m/s (0.75 mph) air motion, and include I_a of 0.8 clo.
[3]Open = without mask, hood, gloves; with open collar, etc.
[4]Closed = with mask, hood, gloves; all apertures closed.
[5]See page 17, paragraph 2.

228

fitted clothing. An example of the inherent variability in
insulation values can be obtained by considering that the
long-sleeved shirt and trousers value of 1.41 clo might go
as low as 1.35 clo for a tight-fitting ensemble, and up to
about 1.43 for a fairly loose-fitting shirt and trousers.

From these changes in insulation value with clothing
fit, obviously any values in the table beyond the first
decimal place are more indicative than precise. An average
man (i.e., surface area of $1.8m^2$) would lose 7.1
kcal/hr.°C difference between skin and ambient air
temperature with 1.4 clo of insulation, but only 5 kcal/hr.
with 2 clo of insulation. A long-sleeved shirt and trousers
provide intrinsic insulation of 0.6 clo and the external air
layer about 0.8 clo in still air. The total of 1.4 clo
results in a heat transfer, for a standard ($1.8m^2$) man of
7.1 kcal/hr.°C. Belding [4] suggested a radiant heat
transfer of 6.6 kcal/hr.°C with such clothing and a
convective heat transfer of $7(WV)^{0.6}$ kcal/hr°C in still
air (e.g., WV = 0.11 m/s or 0.25 mph). This convective
exchange would be 1.8 kcal/hr.°C, providing a combined
transfer of 8.4 kcal/hr°C (i.e., 6.6 + 1.8) by radiation and
convection. With a hot skin temperature [36°C (97°F)] and
environmental temperatures of concern generally in excess of
20°C (68°F), the 15-20 kcal difference in heat loss per hour
resulting from even a 0.5 clo difference in the insulation
of an ensemble is clearly of minor importance; however, as
discussed below, the effect of insulation is much more
substantial in the extent to which it can affect evaporative
heat transfer.

The evaporative cooling allowed by the environment, as
discussed above, is determined from a psychrometric wet bulb
thermometer. Woodcock used this in defining a moisture
permeability index (i_m) for materials; Goldman [7,29]
subsequently applied this concept to measuring and
calculating the maximum evaporative cooling allowed by a
clothing ensemble. The permeability index (i_m) is simply
the dimensionless ratio of the evaporative cooling allowed
by the clothing and its surface air film (i.e., through
$I_{cl} + I_a$), to the maximum evaporative cooling obtainable
by a psychrometric wet bulb thermometer.

Typical permeability index values for most clothing or
materials average about 0.4 (at 0.3m/s air velocity) unless
impermeable layers or water repellent treatments are
incorporated within the clothing assemblies, but the
measured value depends in part on the insulation of the
material or ensemble. Increases in insulation tend to be
matched by increases in measured moisture permeability.
Since impermeable materials tend to be relatively thin,
covering the body surface with these materials may add only
slightly to the total insulation, but will directly reduce

the permeability in a linear ratio to the area covered by impermeable material [23]. Adding an impermeable layer, such as a plastic hood or mask to cover an area of previously exposed bare skin [28], will produce a much more serious reduction in the overall permeability index than covering an equivalent area of the body that is already covered with clothing [15].

The permeability index is a measure of the evaporative characteristics of the clothing materials and associated trapped air layers, but it does not provide a true measure of the evaporative cooling potential from the skin to the ambient environment. The reason i_m is generally fairly constant, at about 0.4 in still air, is that i_m really represents a characteristic moisture diffusion constant through air. The actual effective evaporative cooling obtainable by the clothing wearer is a function of this diffusion constant (i_m) (as modified by unique or water-repellent treatments, very tight weaves, or specifically introduced impermeability), divided by some expression of the length or thickness of the diffusion path; the clo insulation value provides a suitable measure of this thickness. Thus, the net evaporative cooling obtainable by the wearer of a garment system is determined by the permeability index ratio i_m/clo.

Table I includes estimated values of the permeability index (i_m) and the more critical, permeability index ratio (i_m/clo) for a series of clothing assemblies that might be worn for chemical protection. Using the 2.2°C/mmHg Lewis relationship, to link the evaporative heat transfer to the convective heat transfer, it is clear that, with one clo of insulation, one should obtain 22 kcal/hr. (i.e., 10 kcal/hr °C times 2.2°C/mmHg) of heat transfer per mmHg difference between skin (P_s) and ambient air vapor pressure ($\phi_a P_a$) for a standard 1.8 m^2 adult male, if he behaved like a psychrometric wet bulb. Multiplying this potential maximum cooling of 22 kcal/hr.mmHg by the i_m/clo ratio determines the actual potential maximum evaporative cooling allowed through a clothing assembly. Therefore, each change of 0.1 i_m/clo produces a change of 2.2 kcal/hr.mmHg difference between skin and air vapor pressures. The vapor pressure of a hot (36°C) sweaty skin is about 44 mmHg and, at 25°C (77°F), 50% relative humidity, the ambient vapor pressure is about 12 mmHg; under that environmental condition, a change of 0.1 i_m/clo represents a change of approximately 70 kcal/hr. in the maximum evaporative cooling allowed by the ensemble [i.e., 22 x (44-12) x .1]. This direct role played by increasing insulation in the evaporative cooling obtainable by the wearer of a clothing assembly explains why individuals wearing multiple layers of heavy clothing [i.e., high insulation (clo) values] in the winter can become heat casualties during heavy work despite cool-to-cold and relatively dry ambient environmental conditions; they simply cannot get enough evaporative cooling at the skin, despite the very low ambient vapor pressure, to balance their heat production. Note that although Belding appears to have been unaware of the Lewis relationship (2.2°C/mm Hg), the evaporative heat transfer he suggested of 23 $(WV)^{0.6}$ per mm Hg difference between skin and ambient vapor pressure for a long-sleeved shirt and trouser ensemble is 1.9 times the 12 kcal/hr.°C he used for its convective exchange.

The role played by wind speed in altering insulation has been described above, as has the role played by body motion in inducing an effective wind [44]. Body motion has a direct effect on clothing insulation and, hence, effective evaporative cooling (i_m/clo) by "pumping" (i.e., exchanging) the air trapped within the clothing fabric and between the clothing layers. This increases the evaporative and non-evaporative heat exchange with the ambient air. Belding described a reduction of almost 50% in the total insulation of a heavy Arctic ensemble as the wearers went from rest to walking at 3.5 mph on a treadmill. Givoni and Goldman [22] have developed a family of pumping coefficients to characterize the changes in various clothing ensembles with "effective wind" (WV_{eff}), where WV_{eff} was defined as the sum of the ambient air motion and 4% of the increase in M (in Watts) above the resting heat production; i.e., $WV_{eff} = WV + 0.04 (M-105)$. Although, obviously, the appropriate m/s units for air velocity cannot be rationally derived from this totally empiric estimate, this treatment of effective air motion, and the use of a pumping coefficient to characterize the changes in both the clothing insulation and the permeability index ratio, has been demonstrated to be more than adequate (GEGU) to characterize the changes in insulation and permeability of clothing during wearer activity. The pumping coefficient (p) for insulation is the slope of the line connecting two measurements of insulation at different windspeeds on a logarithmic plot; since insulation decreases with increasing wind speed, the pumping coefficient for insulation has a negative exponent. Similarly, the pumping coefficient for the permeability index (i_m) is the slope of the line connecting two determinations of i_m at different effective wind velocities; since permeability increases with increasing wind speed or wearer motion, the pumping coefficient for i_m is a positive exponent. Thus, the form of the pumping coefficient for the permeability index ratio is $(i_m/clo)^{2p}$.

A value of 0.25 can be taken as the pumping coefficient for a long-sleeved shirt and trousers, compared with a value of 0.20 for a completely closed, but air-permeable, charcoal-in-foam, chemical-protective ensemble. The pumping coefficient (p) for a heavy butyl garment, normally worn totally closed with mask, hood and gloves, has not yet been measured but could be less than 0.10. In summary, the insulation of chemical protective clothing is largely a function of its thickness, looseness of fit and number of layers, while its permeability is a direct characteristic of the nature of the chemical protection sought.

There are four possible approaches to provide such chemical protection. First, everyday work clothing can be supplemented with specialized gloves, aprons, face shields, etc., if only partial protection is needed against spatter or skin contact. If respiratory protection is required, this can be provided with a filtered mask [55]. If full-body protection is needed, it can take three forms. Heavy, multi-layered ensembles have been impregnated with chemicals which decompose the toxic agents; e.g., chlorocarbon impregnated underwear and outerwear. Such ensembles have some ability to sweat-wet through and thus allow some evaporative cooling, but they may also produce significant skin irritation. Alternatively, charcoal-in-foam overgarments, worn alone or over normal work clothing, can be used; these garments require liquid-repellent surface finishes to minimize local, surface concentration build-up from overwhelming the adsorbent properties of the charcoal in the garment. The charcoal-in-foam systems have been adopted by the military since they appear to be the most comfortable of the available chemical protective garments, but they may fail when they are soaked through or overwhelmed by massive surface contamination; also, while they are most comfortable when worn in a high wind because they are air permeable, the rate of air movement across the charcoal could be too rapid to insure adsorption of all the toxic chemicals. The fourth, and most frequent industrial choice, is a totally impermeable clothing system. The major drawback to the totally impermeable systems is the high potential of heat stress associated with totally blocking evaporative cooling from the human body.

Fourteen clothing materials have been evaluated for approximately three hundred chemicals, and recommendations for selection among them have been published in "Guidelines for the Selection of Chemical Protective Clothing." (See references in Chapter 9, Personal Protective Equipment.) Chemical protective clothing, considered as a subcategory of personal protective clothing, has been divided into five classifications in these Guidelines:

    a. Head, face and eye protection, which encompasses hoods, face shields and goggles;

b. Hand and arm protection as provided by gloves and sleeves;

c. Footwear protection includes specialized boots and shoe covers;

d. Partial torso protection, provided by an apron, jacket, pants, coat or bib overalls;

e. Complete torso protection, including simple coveralls and full-body encapsulating suits.

A number of approaches to alleviate the heat stress associated with chemical protective clothing are commercially available or under development. These include ice vests and wettable covers, both of which can be extremely effective and simple solutions for the industrial work force, and range to microclimate cooling systems where filtered ambient air, conditioned ambient air, or liquid cooling is supplied within the impermeable ensemble using Vortex tubes, prefrozen (e.g., ice) block heat exchangers, mechanical air conditioners and the like. The potential cooling provided by such systems has been extensively explored in a number of reports [31,51,52,53]. Those available commercially to date appear to require trade-offs between weight carried by the user and limited cooling duration, or require external power and umbilical connections which limit mobility.

ENVIRONMENTAL HEAT STRESS INDICES

Clearly, it would be difficult to consider simultaneously the six separate factors that must be measured to assess the heat stress of individuals at rest or work in a given clothing ensemble in any environment. Accordingly, over the years a series of environmental indices have been developed to express the interaction of two or more of these six factors. It must be recognized that an environmental index is not a precision statement. Instead, it is a ranging term and, as such, falls in the GEGU category. Extensive reviews of the indices for heat and cold are available in the published literature [5,12,24,32].

1. Direct Indices

Of the four environmental factors cited above as essential to measure in order to determine comfort and/or heat stress, air movement and black globe temperature have little meaning per se as environmental indices. The other two, air temperature and wet bulb temperature, can serve as direct indices, the former for thermal comfort and the latter for heat stress, under a proper set of constraints as to clothing and activity level.

a.   Air Temperature and Thermal Comfort

The simplest index of cold and warm conditions in conventional clothing in a conventional work place is obtained from the air temperature (t_{db}) itself. Given conventional indoor clothing (1.4 clo), air motion [< 0.2m/s (40 fpm)] and humidity [40($\pm$20)%], and air temperature equal to mean radiant temperature, the range of dry bulb temperatures from about 22°C to 25.5°C (72 to 78°F) is generally comfortable for sedentary workers (M = 120 Watts $\pm$ 10%) [14]. Note that, having specified commonly occurring values for five of the six factors, a modest range of values can be assigned to the sixth to delineate a zone of thermal comfort; this "passband" for air temperature for comfort is about 3.5°C (6°F) wide [23]. Of course, increasing heat production moves the comfort air temperature band substantially lower, with each increment of 30 Watts (25 kcal/hr) in heat production requiring a lowering of the comfort band by about 1.7°C (3°F); indeed, as pointed out earlier, heat stress can occur at air temperatures below 0°C, given a high enough metabolic heat production and sufficient insulation.

b.   Wet Bulb Temperature and Heat Stress

Tolerance times and temperature sensations can be plotted directly on a temperature vapor pressure diagram. For cold conditions, the dry bulb temperature per se, appears to control discomfort with little or no adjustment for wet bulb temperature (i.e., humidity). For normally clothed or unclad individuals, the wet bulb temperature per se can serve as a satisfactory index of heat stress. The upper limit for unimpaired performance of most cognitive tasks can be taken as a wet bulb temperature of 30°C (86°F) for both normally clothed and unclothed subjects with air movement ranging from 0.1 to 0.5 m/s (20-100 fpm). Note, however, that for individuals wearing chemical protective clothing, where the limitation on evaporative cooling is

234

imposed by the clothing per se rather than by the ambient vapor pressure, wet bulb temperature is not an appropriate index. In such a case, the ambient dry bulb temperature (t_a) or, making a correction for radiation, the adjusted dry bulb temperature (t_{adjb}) is a better index.

2. Rational Indices

In unusual work situations, e.g., performing under high intensity arc lights on a movie set, MRT per se could serve as a rational index of heat stress, but it appears not to have been used as such.

a. Operative Temperature (t_o)

Mean radiant temperature represents the uniform surface temperature of an imaginary black enclosure. Operative temperature represents the uniform overall temperature of the same enclosure, and encompasses an exchange of heat between the man and his environment, by both radiation and convection, to the same degree as in the actual environment. Operative temperature can be derived from the heat balance equation where one defines a combined (i.e., radiation and convection) heat transfer coefficient (h) as the weighted sum of the heat transfer coefficient by radiation (h_r) and the average heat transfer coefficient by convection (h_c). The operative temperature (t_o) [17] is then derived as:

Eq. 5 $$t_o = (h_r t_r + h_c t_a)/(h_r + h_c)$$

The operative temperature represents a more precise form of the adjusted dry bulb temperature $[t_{adjb} = (t_a + MRT)/2]$; the latter should not really be used when extreme radiant temperatures are involved but, in general, is GEGU. The operative temperature can be used directly in the heat balance equation to calculate the heat exchange by radiation and convection (H_{R+C}) as:

Eq. 6 $$(H_{R+C}) = h(t_o - t_{surf}) = h(t_o - t_{sk}) F_{cl}$$

where: t_{surf} = mean surface temperature of the clothing
t_{sk} = mean skin temperature

235

and F_{cl} is an intrinsic thermal efficiency of the clothing [47]. Note that use of the F_{cl} form of expressing intrinsic clothing thermal efficiency, as used in the ASHRAE Handbook of Fundamentals requires an adjustment for the relative increase in the area of the clothed body surface over that of the unclothed body surface. It also requires the addition of an adjusted insulating surface air film. In general, it seems preferable to express the total insulation of a clothing ensemble as directly measured from a copper man [7] in clo units and substitute t_o for t_a to account for the combined convective and radiative heat exchanges.

b.   Heat Stress Index (HSI)

The Heat Stress Index is one of the most useful indices for evaluation of heat stress, in part because Belding and Hatch provided a table of the physiological and hygienic implications of eight-hour exposures at various HSI (cf. Table II). The Heat Stress Index itself is simply an application of the heat balance equation. It is the ratio of the evaporative heat loss required (E_{req}) for thermal equilibrium, to the maximum evaporation (E_{max}) allowed through the clothing that can be taken up by the environment, as discussed previously. An adjustment is required for the maximum rate of sweating which, for an average man approaches 2-3 liters/hr but this sweat rate cannot be sustained. Indeed, under such maximum strain, heat exhaustion usually occurs in less than an hour. The generally accepted value for a sustainable maximum sweat rate is 1 liter/hr, which represents a potential cooling power of some 700 Watts if all the sweat can be evaporated; i.e., each ml (one ml = one gram) of sweat evaporated produces 0.58 kcal of cooling, but unevaporated sweat provides no cooling, uselessly increasing body dehydration. Figure 2 presents a series of nomograms for a graphic solution of HSI which, as developed, assumes a 35°C skin temperature, and a conventional long-sleeved shirt and trousers ($I = 1.4$ clo; $I_{cl} = 0.6$ clo + $I_A = 0.8$ clo) clothing ensemble; note also (cf. Block C of Fig. 3) that E_{max} is limited to 700 Watts. Although informative, it should not be used for clothing other than ordinary, indoor work clothing (i.e., long sleeved shirt and trousers).

236

Table II. Physiological and Hygienic Implications of 8-hour
Exposures to Various Heat Stresses

Index of Heat Stress (HSI)	Heat Stresses
-20 -10	Mild cold strain. This condition frequently exists in areas where men recover from exposure to heat.
0	No thermal strain.
+10 20 30	Mild to moderate heat strain. Where a job involves higher intellectual functions, dexterity, or alertness, subtle to substantial decrements in performance may be expected. In performance of heavy physical work, little decrement expected unless ability of individuals to perform such work under no thermal stress is marginal.
40 50 60	Severe heat strain, involving a threat to health unless men are physically fit. Break-in period required for men not previously acclimatized. Some decrement in performance of physical work is to be expected. Medical selection of personnel desirable because these conditions are unsuitable for those with cardiovascular or respiratory impairment or with chronic dermatitis. These working conditions are also unsuitable for activities requiring sustained mental effort.
70 80 90	Very severe heat strain. Only a small percentage of the population may be expected to qualify for this work. Personnel should be selected (a) by medical examination, and (b) by trial on the job (after acclimatization). Special measures are needed to assure adequate water and salt intake. Amelioration of working conditions by any feasible means is desired, and may be expected to decrease the health hazard while increasing efficiency on the job. Slight "indisposition" which in most jobs would be insufficient to affect performance may render workers unfit for this exposure.
100	The maximum strain tolerated daily by fit, acclimatized young men.

Adapted from Belding and Hatch, "Index for Evaluating Heat
Stress in Terms of Resulting Physiologic Strains," Heating,
Piping, and Air Conditioning, 27:129-136, 1955.

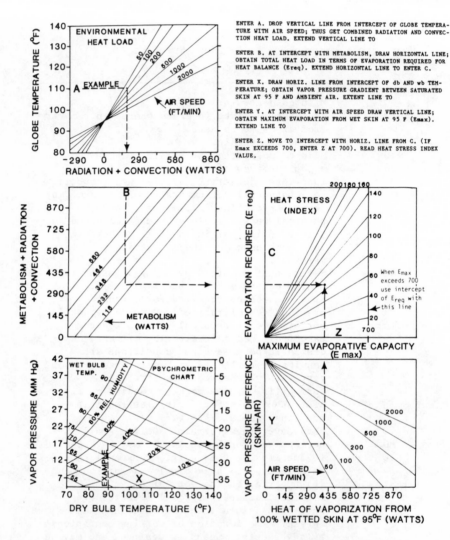

ENTER A. DROP VERTICAL LINE FROM INTERCEPT OF GLOBE TEMPERA-
TURE WITH AIR SPEED; THUS GET COMBINED RADIATION AND CONVEC-
TION HEAT LOAD. EXTEND VERTICAL LINE TO

ENTER B. AT INTERCEPT WITH METABOLISM, DRAW HORIZONTAL LINE;
OBTAIN TOTAL HEAT LOAD IN TERMS OF EVAPORATION REQUIRED FOR
HEAT BALANCE (Ereq). EXTEND HORIZONTAL LINE TO ENTER C.

ENTER X. DRAW HORIZ. LINE FROM INTERCEPT OF db AND wb TEM-
PERATURE; OBTAIN VAPOR PRESSURE GRADIENT BETWEEN SATURATED
SKIN AT 95 F AND AMBIENT AIR. EXTENT LINE TO

ENTER Y. AT INTERCEPT WITH AIR SPEED DRAW VERTICAL LINE;
OBTAIN MAXIMUM EVAPORATION FROM WET SKIN AT 95 F (Emax).
EXTEND LINE TO

ENTER Z. MOVE TO INTERCEPT WITH HORIZ. LINE FROM C. (IF
Emax EXCEEDS 700, ENTER Z AT 700). READ HEAT STRESS INDEX
VALUE.

Figure 2. Nomograms for graphic solution of the
heat balance equation.

238

c.   Skin Wettedness (%SWA)

Percent skin-wettedness, as defined by Gagge [16], is essentially identical to HSI except that, in theory, the percent SWA uses the observed skin humidity or wettedness, rather than the required evaporative cooling as the numerator in taking the ratio to the maximum evaporative cooling power of the environment. Skin wettedness (alternatively, skin relative humidity or the percent of skin surface that is sweat-wetted) appears to be what the body uses to sense its thermal discomfort. There is little, if any, sensory input from deep-body temperature or from skin temperature, although both deep-body temperature and local skin temperature provide the control inputs for regulation of sweating. A worker will generally not continue work which results in skin-wettedness much above a 60% level [%SWA=HSI = 60 or, equivalently, a relative humidity of the skin $(\phi_s) \geq 60\%$ Ps]. At this 60% level, sweat frequently drips from the skin and begins to be wasted, except under conditions of low ambient vapor pressure, minimal clothing and/or reasonably high air movement.

3.   Empirical Indices

a.   Effective Temperature (ET, CET and ET*):

The best known and most widely used of the environmental indices is the effective temperature (ET) index originally derived in 1923 for ASHRAE (35). It is generally calculated from the nomogram given in Figure 3, and combines the effects of dry bulb and wet bulb temperatures and air movement. Substituting the black globe temperature (t_g) directly in place of the air temperature produces a Corrected Effective Temperature (CET) accounting for radiation. Thus, the CET index combines all four of the key environmental factors into a single number. ET and CET were derived using subjective judgments of equivalence by a limited number of subjects. Gagge recently developed a new effective temperature, ET* which uses a 50% r.h. as the reference humidity. The ET* corresponds more closely to familiar sensations at a t_a = ET* than it does to the 100% r.h. referenced ET.

239

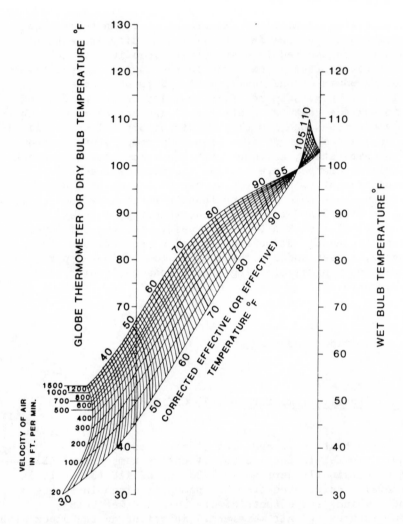

Figure 3.　Chart showing normal scale of corrected effective temperature.

Effective Temperature has been used most extensively for studies of psychological tolerance limiting conditions. It serves as the most useful guideline to the efficiency of a work-force. An ET (CET) greater than 30°C (86°F) generally is considered unacceptable and usually decreases productivity in an industrial work force. A World Health Organization scientific group has proposed tolerance limits to heat stress in terms of ET/CET [61]. The suggested limit for unacclimatized individuals doing sedentary to light work (<215 watts) was 30°C (86°F); for moderate work (to 360 watts) the suggested ET/CET limit was 28°C (82.5°F), and for heavy work (to 500 watts) a limit of 26.5°C (80°F). Fully heat-acclimated individuals (7 days of work in the heat for 2 or more hours each day) were supposed to tolerate 2°C higher ET (CET) levels for an eight-hour daily work shift. These proposed values are generally consistent with thermal environmental conditions in deep mining which resulted in stable rectal temperatures (equilibrium rectal temperatures at "safe" levels) in groups of highly acclimatized South African miners. With very large groups of workers, however, some heatstroke still occurred, probably because of the large individual variability in response to heat stress. Individuals of low maximum oxygen uptake (i.e., small body stature or poor physical condition) appear to be particularly susceptible to heat illness [59]. In addition, there was a slight reduction in productivity of these gold mine workers (~5%) beginning at about ET 82°F (27.7°C), which is also the threshold reported for onset of fatal heatstroke during "hard" work.

b. Wet Bulb Globe Temperature (WBGT)

The WBGT Index (60) uses the naturally convected wet bulb as a measure of the environmental stress, rather than the psychrometric wet bulb used in all other indices presented thus far. The natural wet bulb temperature (t_{nwb}) value is taken as 70% of the WGBT; another 20% is contributed by the black globe thermometer temperature (t_g) directly, and 10% by the dry bulb temperature (t_{db}):

Eq. 7 $$WBGT = 0.7\ t_{nwb} + 0.2\ t_g + 0.1\ t_{db})$$

The WBGT index thus combines the effects of humidity and air movement (in t_{nwb}), and low temperature radiant heat and solar radiation (in t_g), and air temperature (t_{db}). WBGT instruments are commercially available from a number of manufacturers, but some misrepresent the WBGT, using a psychrometric rather than the natural wet bulb; others use smaller globes or unusual radiant heat sensors, with little validation that the adjustment (if any) made in calculating their WBGT is acceptable. Most, being battery operated and requiring calibration, can fail or be easily miscalibrated and the commercial instruments for WBGT, therefore, seem better suited for laboratory than field use. The military still use WBGT guidance for prevention of heat illness during training but it is being supplanted by a new wet globe thermometer (WGT, see below) for operational use.

As pointed out above, for individuals wearing impermeable or reduced permeability clothing, the WBGT is probably not as good an index as the adjusted dry bulb temperature alone. However, for individual workers wearing conventional clothing, WGBT remains the index of choice for expressing physiological tolerance limits at work or rest. The WBGT index was developed by Yaglou [60] to help reduce the number of heat casualties incurred during Marine training in southern U.S. military bases. Its introduction was followed by a dramatic reduction in the incidence of heat casualties when the following guidelines were mandated:

AT WBGT OF: PROCEDURE:

82°F   unseasoned personnel should have only
           limited heavy exercise.
85°F   strenuous exercise such as marching at
           standard rate, should be suspended during
           the first three weeks of unacclimatized
           troop training; outdoor classes in the sun
           should be avoided.
88°F   strenuous exercise should be curtailed for
           all recruits and trainees with less than
           twelve weeks of training in hot weather.
88-90°F   thoroughly conditioned troops, after
           acclimatization each season, could carry
           on limited activities for up to six hours
           per day.

The successful reduction of heat casualties in the military community by adherence to these WBGT guidelines was not lost upon the civilian community. Brief and Confer [62] proposed WBGT limits of 32.2°C (90°F) for light work, 30°C (86°F) for moderate work, and 26.7°C (80°F) for heavy work under indoor situations in 1971.

In 1974, an Advisory Committee on Heat Stress sponsored by NIOSH developed the following table (Table III) on threshold WBGT at which work should be suspended [63,64,65]. Note that the Committee split WBGT levels into

Table III. Threshold WBGT Values Proposed by the Standards Advisory Committee on Heat Stress, 1974 [63,64,65]

Work Load	WBGT in °C Air Velocity:	
	1.5 m/s	1.5m/s
> 300 kcal/hr (>200 W/m^2)	26.1	28.9
201 - 300 kcal/hr (135 W/m^2 - 200 W/m^2)	27.1	30.6
200 kcal/hr (135 W/m^2)	30.0	32.2

Table IV. ACGIH Permissible Heat Exposure Threshold Limit Values in °C WBGT (taken from reference 1)

Work-Rest Regimen	Work Load		
	Light	Moderate	Heavy
Continuous work	30.0	26.7	25.0
75% work-- 25% rest, each hour	30.6	28.0	25.9
50% work-- 50%rest, each hour	31.4	29.4	27.9
25% work-- 75% rest, each hour	32.2	31.1	30.0

two categories as a function of air velocity, with threshold WBGT levels at air velocities of less than 1.5 m/s (300 fpm or 3.3 mph) some 2 to 3°C below the corresponding thresholds for air velocities greater than 1.5 m/s. These WBGT threshold values are not strikingly different from those proposed by Brief and Confer for light and moderate work, if one uses the values for air velocities greater than 1.5 m/s, but are rather different for the heavy work situation. In many areas of the United States, such high WBGT's occur sufficiently often that shutting down would make the plant unprofitable, and infrequently enough that plant air conditioning was not seen as cost-effective. As a result, to date no heat stress standard has been promulgated. Perhaps the next national heat stress standard proposal [38] should follow the WBGT guidelines recommended by the American Commission of Government Industrial Hygienists for permissible heat exposure threshold limit values, given in Table IV [1]. Such recommendations for work-rest cycle alteration should be much more acceptable to both management and labor. Essential work could be continued under quite severe WBGT conditions, albeit for only a limited period during each hour; e.g., at a WBGT of 30°C, heavy work could be performed for fifteen minutes of each hour, with the rest of the hour spent at rest. This would cater to those plants or regions where the occurrence of high WBGT's is infrequent, or of only a few hours duration each afternoon. No investment in major air treatment programs would be required, but there would not be a total loss of productivity, or cessation of all essential activities [25]. Industry would also have a more rational decision basis for recognizing the trade-offs between productivity losses and the economic costs of providing increased ventilation, dehumidification or frank air conditioning for indoor situations. Outdoors, fewer options are available; shading the working area is frequently the most feasible option.

c. Wet Globe Temperature (WGT)

In addition to the problem of reading and mathematically manipulating three temperatures, the WBGT apparatus tend to be set up at a fixed location, frequently quite remote from the working environment. Botsford developed a simpler device, the "Wet Globe Thermometer" (WGT or "Botsball") for use in the aluminum industry. This device is simply a three-inch globe with a black, wettable cover; a standard metallic stem thermometer is inserted into the globe through an extended neck, which contains a small water reservoir to maintain the black cover at 100% wettedness.

The device is simple, portable, and easy to read. A single number value is directly provided, rather than the three separate values provided by the WBGT. Goldman modified the WGT by color-coding the critical zones [48] to provide operational guidance during actual field operations. The WBGT is still in use. The simpler, color-coded WGT has been used in the field with remarkable success. Table V suggest Doctrine on Adjustment of Work-Rest Cycles and Increasing Water Intake. Note the guidance that the Botsball WGT's are equivalent to 2°F higher WBGT's; this 2°F offset value was obtained in several laboratory and field evaluations [31], but will not be correct in all cases. Furthermore, it may be difficult to maintain the WGT surface wet enough in a hot day environment and the offset from WBGT may be much greater than 2°F; more experience is needed. Note also that, under "GREEN" conditions (i.e., WGT between 80-83°F) water intake of between one-half and one quart per hour is recommended for 50-minute work/10-minute break cycles. With increasing heat stress, recommended water intake is further increased and work-rest cycles are decreased.

HEAT STRESS AND PRODUCTIVITY

There have been a great many studies [2,8,39,40,41,42] attempting to relate the productivity of a work force to the environmental heat stress. Most of these have used the effective temperature as an expression of the environmental heat stress, but very few have adequately controlled such key factors as motivation, need or expectancy. Thus, the results of studies on the effects of heat stress on productivity have varied widely. Some actually suggest improved performance under conditions of heat stress when men are in total chemical protective encapsulation. In other studies, very mild heat stress has been shown to decrement performance in men wearing normal work clothing. In response to a report by Fox et al. [11] that increasing heat stress actually improved target detection, Colquhoun and Goldman [11] evaluated the ability to detect a target as a function of increasing body heat storage. Performance in their study involved not only detection of a target, but also a judgment on the certainty with which the target was detected. The results showed that while total target detection did improve, as Fox et al. [11] had stated, the

Table V. Water Intake, Work/Rest Cycles for Essential Field Operations (which cannot be curtailed) for Heat Acclimated Fit Workers

Heat Condition	Botsball WGT (°F)*	Water Intake (qt/hr)	Work/Rest Cycles (Min)
Green	80°-83°	0.5-1.0	50/10
Yellow	83°-86°	1.0-1.5	45/15
Red	86°-90°	1.5-2.0	30/30
Black	90°&above	2.0	20/40**

*To convert WGT to WBGT add 2°F. Below 80° drink up to 0.5 qt/hr, 50/10 work/rest cycles.
**Depending on condition of the worker.

To maintain physical performance:

1. Drink 1 qt. of water in the am, at each meal, and before any hard work.
2. Take frequent drinks, since they are more effective than all at once. Larger workers need more water.
3. Replace salt loss by eating 3 meals per day.
4. As the WGT increases, rest periods must be more frequent, work rate lowered, and loads reduced.
5. Use Water as a key element to maintain top efficiency by drinking each hour.

Table VI. Safe "Closed" Suit Times for Moderate Work (300 w; 250 kcal/hr

Ambient Air (t_a) Temperature (°F)	Wearing Time (Closed)
30° or less	8 hours
30° - 50°	5 hours
50° - 60°	3 hours
60° - 70°	2 hours
70° - 80°	90 minutes
80° - 85°	60 minutes
85° - 90°	30 minutes
90° or above	15 minutes

identification and decision-making skills of the subjects decreased. While more targets were being detected, more non-targets were erroneously identified, with increasing certainty that they were targets as body heat storage increased. One of the most massive studies on the effects of mild heat (and cold) stress on performance has just been reported by Wyon's group. Again, the interaction between environmental discomfort and performance is far from clear, with some tasks decremented and others enhanced in the various subpopulations (males and females, native and caucasian workers) studied.

Decrements in performance are clearly task-dependent, as well as highly dependent on the motivation of the work force. Tasks involving decision making, judgment and complex mental functions appear to decrement at much greater rates than rote tasks such as addition. Physical task performance is relatively insensitive until the workers are affected by incipient heat exhaustion. The ability of two journeymen electricians to install duplex outlets was essentially unimpaired as Effective Temperature was increased from 21 to 27°C (70 to 80°F) across a full spectrum of relative humidities, but was reduced by 10% at 32.2°C (90°F) ET with the greatest reductions at the highest relative humidities. At 38°C (100°F) ET, productivity ranged from a low of 57% at 90% r.h., to 84% for r.h.< 40%, while at 43°C (110°F) ET, productivity was negligible above 80% RH, and ranged from 50-60% at relative humidities between 20-70%. This limited data base on a few highly trained and extremely well-motivated workers indicates that the decrements for such rote tasks can be relatively small under most heat stress conditions. In contrast, the National Association of Building Contractors suggests, as guidelines for cost estimating, that productivity for tasks involving gross motor skills will be decremented by about 30% at 27°C (80°F) ET, 40% at 32.2°C (90°F) ET, and 60% at 38°C (100°F) ET. Since these figures are used in estimating costs (and no cost estimator intends to lose money on contract bids), these estimates probably overstate the losses in productivity. Again this data base applies to normally dressed individuals, and not to individuals wearing chemical protective garments.

GUIDELINES FOR HEAT STRESS IN PROTECTIVE CLOTHING

1. Inadequacy of WBGT and WGT

Both WBGT and WGT depend primarily ($\geq$70%) on the natural evaporation allowed by the environment. This would suggest that WBGT and WGT, while perhaps the best heat stress indices for individuals wearing normal clothing (even, perhaps, with some partial, chemical protective, impermeable coverage such as aprons, masks or gloves) become increasingly inappropriate guides as one moves to the reduced permeability charcoal-in-foam overgarments, unless a very stiff breeze is blowing. WBGT and WGT values are probably quite inappropriate to use as guidance for workers encapsulated in totally impermeable chemical-protective clothing. It is important to note, however, that at upper levels of environmental heat stress [except in desert (hot-dry) environments] all these environmental indices give somewhat similar values; i.e., C.E.T., WBGT and WGT all tend to have similar values.

2. Variability in Heat Tolerance Between Groups

Most of the well-known environmental heat stress data base was generated on physically fit, highly motivated young men in military studies, or the select population of fit young mine workers, pre-screened to eliminate individuals with lower heat tolerance, in South Africa [59]. Witherspoon and Goldman have reviewed work rate and ET interactions from large number of military and civilian data bases and produced Figure 4, which includes one line for significant discomfort or change in deep-body temperature for a mixed population and another line for maximum equilibrium (i.e., acceptable, steady state) values representing tolerance for at least four hours without collapse in a highly fit, young population. The data points are plotted as a function of work rate across a range of ET (or CET) values and include data from Africa, India, Germany, U.K. and U.S., and include clothed and unacclimatized populations as well as unclothed, usually acclimatized, populations. The consistency of the findings is perhaps the most remarkable thing about this diverse data base. The same authors point out the relative uniformity in heat tolerance between individuals of comparable age and fitness, as shown in Figure 5, where the data from seven

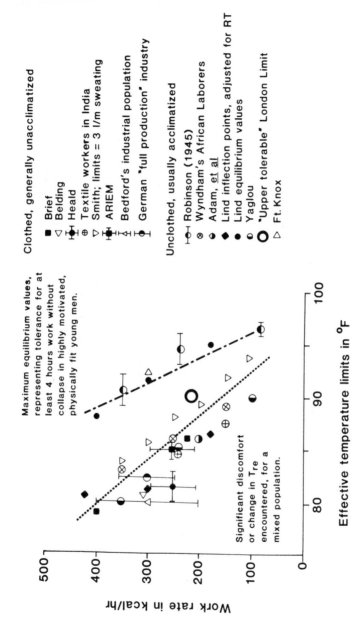

Figure 4. Work rate limits as a function of effective temperature.

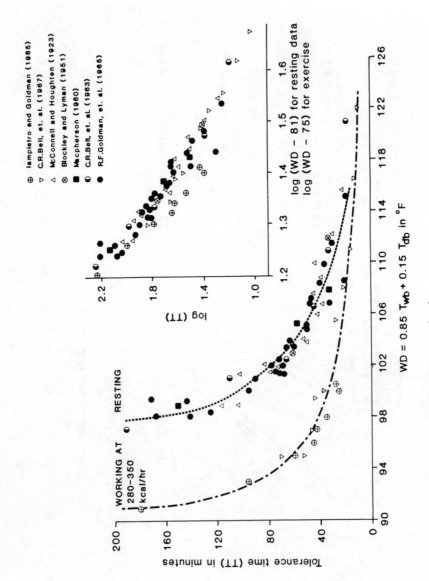

Figure 5. Tolerance time limits (TT).

250

studies between 1923 and 1967 are presented for fit, young men, wearing minimal clothing (usually shorts and boots), at work (280-350 kcal/hr) or at rest, in a range of very hot environments; the environmental heat stress is expressed as the Oxford Index WD (= $0.85 \ t_{wb} + 0.15 \ t_{db}$), using the psychrometric wet bulb and tolerance time is expressed in minutes. Note that even under the most severe conditions, average tolerance time is twenty minutes; as long as the pain threshold at the skin surface is not reached [a skin temperature of about 45°C (113°F)], some twenty minutes of time was provided simply by "mass damping" before body temperatures or heart rates reached critical levels [in these studies, 39.2°C (102.5°F) and 180 b/min, respectively [30,37]]. As shown by the inset, when one compiles all the data from these seven studies on a log-log plot (adjusted to a base of 75°F for the 300 kcal/hr exercise data and to a base of 81°F at rest) the correlation coefficient is extremely high (r~0.96). This implies that about 92% (i.e., r^2) of the tolerance time can be explained simply by the environmental heat stress as expressed by this WD index. Under conditions of severe heat stress, individual variability appears to be minimal within populations of comparable fitness. In an industrial population, however, individuals vary substantially with respect to their state of heat acclimatization, body size and fitness, degree of hydration, congenital sweat gland distribution, etc. Thus, although Figure 4 suggests remarkable agreement in heat stress response across quite divergent populations, appropriate screening is suggested if generalized guidelines are to be used for an industrial workforce required to wear complete chemical protective clothing ensembles. Wear of totally impermeable clothing will, of course, further homogenize the responses of a population by wiping out any differences in effective sweating and differences in heat acclimatization status, other than perhaps those associated with more rapid dehydration in well heat-acclimated individuals. Thus, given an equivalently screened population, it seems safe to conclude that conditions which are sufficiently heat stressful to produce problems for one or two workers are not far from being unreasonably stressful for the total population.

3.   Impermeable Protective Clothing Guidelines

By now, it should be obvious that almost all the indices discussed have limited applicability to men in totally impermeable clothing [43]. The ambient vapor pressure is a meaningless measure of the environmental heat stress for individuals wearing impermeable clothing, as are all indices in which ambient wet bulb temperature (psychrometric or non-psychrometric) is a major factor. The dry bulb (air) temperature, adjusted for solar heat load by using either the adjusted dry bulb temperature, or by an increment for solar radiation of 7°C (13°F) times the percent cloud cover is probably the most appropriate index for totally impermeable garments. Custance has collated data (cf. 29) (Table VI) for safe "closed" impermeable suit times for moderate work (250 kcal/hr) from six U.S., Canadian and Russian studies, as a function of air temperature. As indicated below, predictive modeling has reached the stage where quite valid predictions and extrapolations can be made [27]. For specific cases, predictive modeling should provide the best guidelines for comfort, discomfort, performance or tolerance limits, and risk.

AGE, GENDER AND HEAT STRESS

The effects of age in predisposing to heat stress appear to be primarily associated with the cardiovascular system, although older individuals start to sweat later, and produce less sweat during comparable work in the heat than younger individuals. While older individuals appear to have a higher peripheral blood flow during work in the heat, their maximal work capacity (i.e., $\dot{V}O_{2max}$) is almost always reduced. Accordingly, even healthy, older individuals tend to have reduced heat tolerance to work, exhibit higher heart rates and slightly higher deep-body temperatures, and take longer to return to normal body temperatures. The effects of heat stress on individuals with specific medical problems (e.g., cardiorespiratory difficulties) can, of course, be devastating [9,13].

The original proposed Guidelines for an OSHA Heat Standard [63] postulated differences between male and female workers with respect to their tolerance to heat. Such recommendations were based on the limited research data base available prior to 1974 [25], which suggested that females, as a group, were less tolerant than males to heat stress. Based on newer studies, however, it appears that gender-free standards can be established; brought to equivalent levels of fitness and heat acclimatization, the remaining difference between males and females of any real significance in respect to heat resistance, is body size. A smaller, less fit male is at no less or greater risk to heat

252

stress than an equivalent-sized, less fit female while, with increasing fitness or greater opportunities for heat acclimatization, females and males appear to perform equally well in the heat.

BODY RESPONSES TO HEAT STRESS AS EXPOSURE LIMITS

By now, it should be clear that heat stress simply means: 1) heat losses by radiation and convection are less than the heat produced by metabolism so that a requirement for sweat evaporative cooling exists; and 2) the degree of heat stress is a function of the extent to which the requirement for evaporative cooling can be met. As indicated above, the ratio E_{req}/E_{max} is perhaps the best single indicator of heat stress. A chart (modified from one originally developed by Belding) delineating the physiological responses to heat stress is presented as Figure 6. These responses include, as a first line of defense, an initial rise in skin temperature as a result of vasodilation. The increased blood flow to the skin raises skin temperature and thus helps increase heat loss by convection and radiation, or reduce heat gain when the operative temperature is higher than skin temperature. This first line of defense is limited however, as the available circulating heat transfer fluid of the body (i.e., the blood) becomes depleted by sweating in the absence of adequate rehydration, or becomes pooled in the periphery in the absence of continued muscle activity massaging the venous blood back, past the valves in the veins, to the heart to provide adequate venous return for continued blood flow to the brain. Inadequate blood flow to the brain leads to heat exhaustion collapse. The transition from the vertical, upright posture during work, to the horizontal, recumbent posture of heat exhaustion collapse tends to solve the problem of inadequate venous return, but at an unacceptable cost. Heat exhaustion is most likely to occur in working individuals when, upon suddenly stopping work, they incur massive peripheral pooling and inadequate venous return, and suffer the consequent blackout. It can also occur in individuals while working. Inadequate venous return to the heart leads to decreased central blood pressure and blood flow to the brain, so a signal to "beat faster" is sent to the heart. As in most pumps, sufficient filling time must be allowed between strokes and, as heart rates exceed 180/min, inadequate filling time leads to further decreases in cardiac output with a resulting transition from the vertical to the horizontal state of the worker.

253

Obviously, there is a general, albeit complex relationship between body temperature and heart rate [58]. In general, an increase in working heart rate of more than 30 b/min above the resting level may be considered unacceptable for an industrial work force. Performance decrements may be expected if sustained heart rates exceed 100 b/min for 8 hours, or perhaps 120 b/min for 8 hours if the workers are extremely fit and well heat-acclimatized. For very fit and acclimatized young workers, sustained heart rates of 140 b/min may be compatible with 4 hours of work; 160 b/min may be compatible with 2 hours of work, but heart rates above 180-190 b/min are generally considered unacceptable for any sustained period. Such values, however, are only appropriate for a relatively young work force. The maximum heart rate for individuals can be characterized as a function of age by the relationship: maximum heart rate equals 220 b/min minus age in years. A more appropriate generalization, then, would be that heart rate increases of less than 20% of the difference between an individual's age adjusted, maximum heart rate and his resting level are quite reasonable. For example, the predicted maximum heart rate for a 60-year old is 160 b/min and, if the resting rate is 70 b/min, then the working heart rate should be $\leq$ 88 b/min, i.e., [70 + 0.2 (160-70)]. As general guidelines, increases of 40% of the difference between this age adjusted maximum and the resting level probably represent an uncomfortable level of work; for the 60-year old above, this would be 106 b/min, while for a 20-year old it would be 122 b/min, given a 70 b/min resting value for both. Performance decrements may be expected at heart rates representing 40-60% of this "heart rate increase capacity," tolerance time limits will generally be associated with values between 60 and 80% of capacity, while damage may result at levels requiring more than 80% of an individual's heart rate increase capacity.

The body's second line of defense, sweating, is limited in the cooling it can provide to a sustainable rate of about 700 Watts, even if all the sweat can be evaporated efficiently at the skin surface. Sweating becomes increasingly ineffective as ambient humidities increase so that more sweat is wasted, or as sweat is absorbed into the clothing and evaporation takes place at sites more removed from the skin. Individuals wearing impregnated, but sweat permeable protective clothing frequently receive less than one-half the full cooling benefit of the sweat evaporation that takes place. Since, under conditions of work-associated environmental heat stress, the amount of sweat produced is directly titrated to the amount of sweat evaporation required and obtained, one of the best measures of the role of a clothing ensemble in stressing the wearer is the ratio of the sweat evaporated (E) during a given time period to the maximum sweat produced (P) by the individual. One simply obtains initial and final unclothed weights, adjusted for any water intake, as a measure of the total sweat production (P) and the initial and final clothed body weights as a measure of the total sweat evaporation (E). The E/P ratio with typical clothing ensembles under comfortable conditions will be close to 90%, but will decrease to about 70% with increasing humidity, decreasing air motion, or heavier than normal clothing. The E/P ratios decrease to 40% or less with most "semipermeable", encapsulating chemical protective ensembles, reaching about 20% when these are worn under hot humid conditions or during heavy work. The E/P ratio obviously approaches zero for totally impermeable, encapsulating clothing systems. Again, the relative stress can be divided into the five categories: comfortable, uncomfortable, performance decrementing, tolerance time limiting and perhaps damaging, using the 0-20%, 20-40%, etc., rubric. Thus, and E to P ratio between 80 and 100% will be quite comfortable, between 60 and 80% may be uncomfortable, between 40 and 60% will probably be performance decrementing, between 20 and 40% will be tolerance limiting, and an E/P ratio of less than 20% associated with high risk.

A common problem associated with heavy sweating is the induction of dehydration, since thirst is an inadequate stimulus for drinking enough water to prevent dehydration. Up to 2% of the total body weight may represent excess extracellular fluid that can be lost without major decrements in temperature regulation or work performance capacities, although subtle decrements in psychomotor performance may be associated with lower dehydration levels of 1 to 2%. However, individuals given unlimited access to water (albeit perhaps warm and not necessarily very palatable) have been shown [2] to incur "voluntary dehydration" levels of 8 or 9%, with associated major decrements in performance and greatly increased rates of rise of deep-body temperature. Using a 10% dehydration level as the maximum possible, (i.e., acute loss of 10% of body weight by dehydration although survival has been reported at dehydration levels of 18-20%), levels of 0-2% dehydration (judged from body weight loss during a work shift) would be "comfortable", 2-4% would be uncomfortable, 4-6% would be performance decrementing, and levels above 6% would be associated with limited tolerance times, again using the 0-20, 20-40, 40-60%, etc., rubric of demand/capacity effects.

Another heat-associated syndrome which some individuals suffer is "hyperventilation", particularly in response to exposure to hot-wet environments. The overbreathing results

in a reduction of the normal blood carbon dioxide
concentrations. One of the first subjective sensations is a
tingling around the lips and some dizziness. The phenomenon
is associated with a reduction in blood flow to the brain,
because of cerebral vasoconstriction, and can lead to
blackout as well as to a very diagnostic cramping of the
fine muscles of the hand and foot (carpo-pedal spasm).
While not apt to occur in individuals performing hard work,
and thus producing substantial volumes of carbon dioxide, it
can cause collapse in individuals performing light work or
at rest, and is probably a major contributor to the onset of
heat exhaustion collapse in individuals taking a short break
during periods of intensive work. In many experiments on
heat stress, heat exhaustion collapse occurred during the
period when the individual was asked to stop work so that
his heart rate could be measured by palpation at the wrist;
the resultant collapse during the one-to-two minutes
subsequent to cessation of work probably is contributed to
by both inadequate venous return from peripheral pooling,
and reduction of the carbon dioxide levels in the blood due
to hyperventilation. Hyperventilation may be more
pronounced when respirators and fullface masks are worn, and
the restricted visual field of a gas mask may interact with
heat stress to increase dizziness and nausea. The
additional dead space of such respiratory protection may not
adequately compensate for potential hyperventilation.

PREDICTIVE MODELING AND HEAT STRESS GUIDELINES

As suggested previously, predictive modeling may be the
most appropriate approach for establishing realistic
guidelines for individuals or groups wearing chemical
protective clothing. Such models are particularly well
suited to simultaneously treating the possible variations in
degree of protection, and the type of protective clothing
worn (aprons, charcoal-in-foam, air-permeable but
chemical-impermeable, or totally impermeable ensembles worn
open or closed). Such models can also handle simultaneously
such questions as age and gender, as well as body weight,
height, air temperature, solar load, humidity and air
motion, without concern as to whether an index relying on
ambient vapor pressure, or on psychrometric instead of
non-psychrometric wet bulb temperature needs to be used.
One such model [19,20,22] rigorously addresses the physical
heat transfers allowed by the worker's clothing between an
environment and the worker, and can make adjustment for
individual (or group) capacities to meet heat stress,

including degree of heat acclimatization, and extent of
dehydration. Model outputs include predicted subjective
comfort vote (PMV, on a scale of +3 to -3, where 0 is
comfortable, +3 is very hot), deep-body temperature and
heart rate, Watts of required and maximum evaporative
cooling, and percent sweat-wetted skin area and grams of
sweat produced. This model assumes a relatively fixed skin
temperature, in the 35-36°C range. The model has been
programmed to provide tabulated outputs of rectal
temperature and heart rate as a function of time. It can
also provide values of the equilibrium rectal temperatures
and heart rates that the body will attempt to achieve to
establish heat balance, without recognition of whether or
not these required equilibrium levels may be totally
incompatible with tolerance limitations and, thus, may lead
to collapse long before any equilibrium is reached. In this
equilibrium state mode, the output also identifies the
relative contributions of the work level [46] of any
non-evaporative heat transfer limitations associated with
high ambient temperatures or heavy insulation and also
differentiates any problems associated with high ambient
vapor pressures or inadequate permeability [26]. A sample
output showing the changes associated with changing the
insulation of a protective ensemble (from the 1.4 clo value
of a standard long-sleeved shirt and trousers, by $\pm$ 0.2 clo
increments) is presented in Table VII for men, at rest or
working at 250 or 500 Watts at 20°C (68°F), 25% R.H. with
low air movement (0.3m/s). The output can also identify the
effects of lack of (or improved) heat acclimatization and of
limited (or enhanced) water ingestion programs.

A third output format simply graphs the predicted
rectal temperatures and heart rates over time as a function
of any sequence of rest-work-recovery-work-recovery, etc.
cycles of various durations. This allows optimum
relationships between work and rest periods to be set up
either to keep body heat storage below the 80 kcal/hr
associated with unwillingness to continue work, or keep
deep-body temperatures below the 39.2°C (102.5°F) level at
which there is roughly a 25% risk of heat exhaustion
collapse for individuals wearing chemical protective
ensembles. For example, this could occur under conditions in
which skin temperature cannot be reduced as a result of
sweat evaporation with the increased effective wind velocity
during work because of clothing impermeability. Such
graphic outputs can be color-coded to represent acceptable
heart rate and rectal temperatures for a civilian workforce
under OSHA regulations in one color (i.e., T_{re} < 38.2°C;
heart rate < 120 b/m). A different color can be used to
reflect "safe", albeit stressful, conditions for fit young
adults (i.e., T_{re} < 39.2°C; heart rate < 140, or 160, or

Table VII. Predicted Responses to Changes in Clothing
Ta=20C, RH=25%, WV=0.3m/s. Clo=a/s, Im=.43 p=.25

Clo eff	:	PMV	:	Tref	:	REST Ereq	:	Emax	:	% SW	:	SWEAT
.80		-1.5		36.9		-128		545		0		0
1.00		-1.2		37.0		- 82		436		0		0
1.20		-1.0		37.1		- 51		363		0		0
1.40		-0.8		37.2		- 28		311		0		0
1.60		-0.5		37.3		- 12		273		0		0
1.80		-0.3		37.3		- 1		242		1		2
2.00		+0.0		37.4		12		218		5		25

M=250 W

Clo eff	PMV	Tref	Ereq	Emax	% SW	SWEAT
.61	-0.6	37.4	- 56	934	0	0
.76	-0.3	37.5	6	747	1	7
.92	+0.1	37.6	46	622	7	61
1.07	+0.4	37.7	75	533	14	106
1.22	+0.7	37.7	97	467	21	145
1.38	+1.1	37.8	114	415	28	180
1.53	+1.4	37.9	128	373	34	211

M=500 W

Clo eff	PMV	Tref	Ereq	Emax	% SW	SWEAT
.51	+0.2	38.3	131	1365	10	120
.63	+0.6	38.4	205	1092	19	208
.76	+1.1	38.5	254	910	28	280
.88	+1.6	38.6	289	780	37	342
1.01	+2.0	38.7	315	682	46	397
1.14	+2.5	38.8	336	606	55	446
1.26	+2.9	38.9	352	546	65	490

180 b/m depending on duration of the exposure in chemical protective garments). Transition to still another color can indicate conditions of potential heat exhaustion collapse (i.e., T_{re} between 39.2 and 41°C). Finally, another color (or symbol) can be used for temperatures above 41°C where risk of heatstroke exists.

"CONVERGENCE": THE BEST PHYSIOLOGICAL GUIDE TO HEAT LIMITS

Any tolerance limit established simply on deep-body temperature or heart rate will not address the critical problem termed "skin temperature convergence" [49]. Heat exhaustion collapse has occurred in individuals with deep-body temperatures < 38.1°C (100.6°F) with heart rates, measured only minutes before collapse ensued, on the order of 120-130 b/min after some 30 minutes of exposure to work in the heat while wearing impermeable protective garments. The critical event in such cases is that skin temperature rises, converging toward rectal temperature. A core to skin temperature difference of less than 1°C is a strong indication of inability to continue work in the heat very much longer (see Figure 6). As skin temperature converges toward deep-body temperature, as indicated previously, each liter of blood has a reduced capacity for moving heat from the deep-body centers to the skin from which, in normal clothing, it is eliminated. Associated with this convergence is an increased accumulation of blood in the periphery, and increased heart rates in an attempt to maintain heat balance and blood flow to the brain. In recent field trials in troops wearing chemical protective clothing ensembles, voluntary discontinuance occurred almost contemporaneously with skin temperature reaching deep-body temperature, even though deep-body temperature was well below the usual 39.5°C (103°F) used as a limiting criterion for fit, young, heat-acclimatized troops.

Except as an indicator of risk of heat stroke, deep-body temperature should not be used for individuals wearing chemical protective clothing since heat exhaustion collapse can occur at deep-body temperatures close to 38°C. Also, deep-body temperatures frequently continue to rise after cessation of work, particularly in situations with low evaporative cooling potential (e.g., a high vapor pressure inside personal protective clothing), simply because of the lag time of deep-body temperature as measured by rectal temperature. Since heart rates can rise much too rapidly upon convergence of skin and rectal temperatures to be used safely as a criterion for removal of workers from the heat, what physiological end-point, if any, should be used?

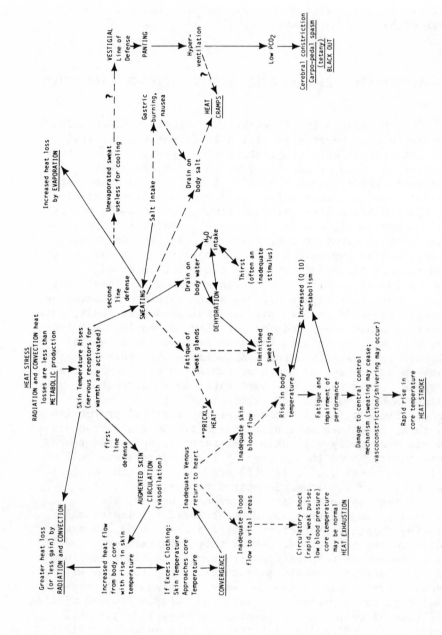

Figure 6. Heat stress and its associated physiologic responses and pathologies

Some years ago, Iampietro and Goldman [37] indicated that skin temperature achieved its equilibrium value very rapidly during work in the heat; the skin temperature at the end of ten minutes of exposure to work in a given hot condition was generally fairly close to the equilibrium skin temperature that would be established, at least until rising deep-body temperatures drove skin temperature up further. They suggested that a reasonable estimate of tolerance time for work in the heat could be obtained based on the skin temperature at ten minutes after onset of work in the heat. Schwartz, subsequently, drew similar conclusions (54). In view of the critical relationship of skin temperature convergence as an end-point for men working in the heat when wearing chemical protective clothing systems, it is recommended that, if any physiological end-points are to be adopted as safety criterion for individuals working in chemical protective clothing, skin temperature should be one of the most important physiological criteria. Skin temperature can be measured at a number of sites, with three sites (chest weighted at 50%, forearm weighted at 14%, and calf weighted at 36%) usually being used to obtain a mean-weighted skin temperature. Under heat stress conditions, however, particularly when wearing complete chemical protective encapsulation systems, skin temperatures become remarkably uniform and measurement of a single skin temperature should suffice. Soule and Goldman have suggested that a lateral or medial thigh temperature is probably the most representative of average skin temperature, and a medial thigh temperature would be least apt to be influenced by direct impingement of solar or other radiant heat sources. A skin temperature at that point (or an average mean weighted skin temperature) in excess of 36°C should be considered prognostic of difficulty in maintaining an acceptable heat balance. A skin temperature above 37°C should be cause for cessation of work in the heat.

REFERENCES

1. ACGIH, American Conference of Governmental Industrial Hygienists, 1980, Threshold Limit Values for Chemical Substances and Physical Agents in the Workroom Environment with Intended Changes for 1980, Cincinnati.
2. Adolph, E. F., and Associates. "Physiology of Man in the Desert." NY: Interscience, 1974.
3. Bedford, T., and C. G. Warner. "The Globe Thermometer in Studies of Heating and Ventilating," J. Hygiene (Camb.) 34:458-473 (1934).
4. Belding, H. S. "Heat Stress," in: Thermobiology, A.H. Rose, Ed. (New York: Academic Press Inc., 1971).
5. Belding, H. S. "The Search for a Universal Heat Stress Index" in: Physiological and Behavioral Temperature Regulation, J. D. Hardy, A. P. Gagge, and J. A. J. Stolwijk, Eds., (Springfield, IL: C C Thomas, Inc., 1970).
6. Belding, H. S. and T. F. Hatch. Index for evaluating heat stress in terms of resulting physiological strain. ASHRAE TRANSACTIONS 62:213-236, 1956.
7. Breckenridge, J. R. and R. F. Goldman. "Effect of Clothing on Bodily Resistance Against Meteorological Stimuli," in: Progress in Biometeorology, S. W. Tromp, Ed. (Lesse The Netherlands: Swets and Zeitlinger, 1977), Vol. 1, Part II: 194-208.
8. Brouha, L. Physiology in Industry. (New York: Pergamon Press, 1960).
9. Burch, G. E., and N. P. DePasquale. Hot Climates, Man and His Heart, (Springfield, IL: C C Thomas Inc., 1962).
10. Colin, J., and Y. Houdas. "Experimental Determination of Coefficient of Heat Exchange by Convection of the Human Body," J. Appl. Physiol. 22:31-38 (1967).
11. Colquhoun, W. P., and R. F. Goldman. "Vigilance Under Induced Hyperthermia," Ergonomics, 15:621-632 (1972).
12. Dukes-Dobos F. N., and A. Henschel. "The Modification of the WBGT Index for Establishing Permissible Heat Exposure Limits in Occupational Work," United States Health, Education and Welfare, National Institute for Occupational Safety and Health, TR-69, (1971).
13. Ellis, F. P. "Mortality from Heat Illness and Heat Aggravated Illness in the United States," Environmental Research, 5:1-58 (1972).
14. Fanger, P. O. Thermal Comfort, (New York: McGraw-Hill, Inc., 1973).
15. Fourt, L., and N. R. S. Hollies. Clothing Comfort and Function, (New York: Marcel Decker Inc., 1970).
16. Gagge, A. P. "A New Physiological Variable Associated

with Sensible and Insensible Perspiration," Am. J. Physiol. 120:277-287 (1937).

17. Gagge, A. P. "Standard-Operative Temperature, a Generalized Temperature Scale, Applicable to Direct and Partitional Calorimetry," Am. J. Physiol. 131:93-103 (1940).

18. Gagge, A. P., C. E. Winslow, and L. P. Herrington. "The Influence of Clothing on Physiological Reactions of the Human Body to Varying Environmental Temperatures. Am. J. Physiol. 124:30-50 (1938).

19. Givoni, B. and R. F. Goldman. "Predicting Effects of Heat Acclimatization on Heart Rate and Rectal Temperature. J. Appl. Physiol. 35:875-979 (1973).

20. Givoni, B. and R. F. Goldman. "Predicting Heart Rate Response to Work, Environment and Clothing," J. Appl. Physiol. 34:201-204 (1973).

21. Givoni, B. and R. F. Goldman. "Predicting Metabolic Energy Cost," J. Appl. Physiol. 30:429-433 (1971).

22. Givoni, B. and R. F. Goldman. "Predicting Rectal Temperature Response to Work, Environment and Clothing," J. Appl. Physiol. 32:812-822 (1972).

23. Goldman, R. F. "Evaluating the effects of clothing on the wearer." Chap. 3, Bioengineering, Thermal Physiology and Comfort, (K. Cena, J. A. Clark, eds.) pp. 41-55, Elsevier, NY, 1981.

24. Goldman, R. F. "Environmental Limits, Their Prescription and Proscription," Intl. J. Environ. Sci. 2:193-204 (1973).

25. Goldman, R. F. "Prediction of Heat Strain Revisited 1979-1980," in: Proceedings of the NIOSH Workshop on the Heat Stress Standard, Cincinnati, September 1979.

26. Goldman, R. F. "Prediction of Human Heat Tolerance," in Environmental Stress, S. J. Follinsbee et al., Eds. (New York: Academic Press, Inc. 1978), pp. 53-69.

27. Goldman, R. F. "Tactical Implications of the Physiological Stress Imposed by Chemical Protective Clothing Systems," in: Proceedings of the 1970 Army Science Conference, (West Point, NY: US Military Academy, 1970).

28. Goldman, R. F. "Tolerance Time for Work in the Heat When Wearing CBR Protective Clothing," Mil. Medicine 128:776-786 (1963).

29. Goldman, R. F. and J. R. Breckenridge. "Current Approaches to Resolving the Physiological Heat Stress Problems Imposed by Chemical Protective Clothing Systems," in: Proceedings of the Army Science Conference, Volume IV, (West Point, NY: US Military Academy, June 1976) pp. 447-453.

30. Goldman, R. F., E. B. Green and P. F. Iampietro. "Tolerance of Hot, Wet Environments by Resting Men," J. Appl. Physiol. 20:271-277 (1965).

31. Goldman, Ralph F. and Staff. "Microclimate Cooling For Combat Vehicle Crewmen," in: Proceedings of the 1982 Army Science Conference, (West Point, NY: US Military Academy, June 1982).

32. Gonzalez, R. R., L. G., Berglund, and A. P. Gagge. "Indices of Thermoregulatory Strain for Moderate Exercise in the Heat," J. Appl. Physiol. 44:889-899 (1978).

33. Haisman, M. F. and R. F. Goldman. "Effect of Terrain on the Energy Cost of Walking with Back Loads and Handcart Loads," J. Appl. Physiol. 36:545-548 (1974).

34. Hardy, J. D. "Heat Transfer," in: Physiology of Heat Regulation and Science of Clothing, L. H. Newburgh, Ed. (London: W.D. Saunders Ltd., 1949) pp. 79-108.

35. Houghten, F. C., and C. P. Yaglou. "Determining Lines of Equal Comfort," ASHRAE TRANSACTIONS 29:163-176, 361-384 (1923).

36. Hughes, A. L., and R. F. Goldman. "Energy Cost of 'Hard Work'," J. Appl. Physiol. 29:570-572 (1970).

37. Iampietro, P. F. and R. F. Goldman. "Tolerance of Men Working in Hot Humid Environments," J. Appl. Physiol. 20:73-76 (1965).

38. ISO, International Organization for Standardization, 1981, Hot environments-determination of the Wet Bulb Globe Temperature (WBGT) Heat Stress Index. Draft International Standard ISO-DIS 7243.

39. Joy, R. J. T. and R. F. Goldman. "A Method of Relating Physiology and Military Performance: A Study of Some Effects of Vapor Barrier Clothing in Hot Climate," Mil. Med. 133:458-470 (1968).

40. Kerslake, D. M. The Stress of Hot Environments. (Oxford, England: Cambridge University Press 1972).

41. Leithead, C. S., and A. P. Lind. Heat Stress and Heat Disorders. (London: Churchill, 1964).

42. MacPherson, R. K., and F. P. Ellis. Physiological Responses to Hot Environment. (London: Medical Research Council, Her Majesty's Stationary Office, 1960).

43. Mihal, C. P. "Effect of Heat Stress on Physiological Factors for Industrial Workers Performing Routine Work and Wearing Impermeable Vapor-Barrier Clothing," Am. Ind. Hyg. Assoc. J. 42:97-103 (1981).

44. Mitchell, D. "Convective Heat Transfer in Man and Other Animals," in: Heat Loss from Animals and Man, J. L. Montieth and L. E. Mount, Eds. (London: Butterworth's 1974).

45. Newburgh, L. H. Physiology of Heat Regulation and the Science of Clothing (Philadelphia, PA: W. B. Saunders, 1949).

46. Nielsen, M. "Die Regulation der Korpertemperatur bei Muskelarbeit," Scand. Arch. Physiol. 79:193-230 (1938).

47. Nishi, Y., and A. P. Gagge. "Moisture Permeation of Clothing: A Factor Covering Thermal Equilibrium and Comfort," ASHRAE TRANSACTIONS 76:1-8 (1970).

48. Onkaram B., L. A. Stroschein and R. F. Goldman. "Three Instruments for Assessment of WBGT and a Comparison with WGT (Botsball)," Am. Ind. Hyg. Assoc. J. 41:634-641, 1980.

49. Pandolf, K. B. and R. F. Goldman. Convergence of skin and rectal temperatures as a criterion for heat tolerance. Aviat. Space Environ. Med. 49:1095-1101 (1978).

50. Passmore, R., and J. V. G. A. Durnin. Energy, Work and Leisure (London: Heinemann Educational Books Ltd., 1967).

51. Raven, P. B., A. Dobson and T. O. Davis. "Stresses Involved in Wearing PVC Supplied-air Suits: A Review," Am. Ind. Hyg. Assoc. J. 40:592-599 (1979).

52. Shapiro, Y. K. B. Pandolf, M. N. Sawka, M. M. Toner, F. R. Winsmann and R. F. Goldman. "Auxiliary Cooling: Comparison of Air-Cooled Versus Water-Cooled Vest in Hot-Dry and Hot-Wet Environments," Aviat. Space and Environ. Med. 53:785-789 (1982).

53. Shvartz, E. "Effect of Neck Versus Chest Cooling on Responses to Work in Heat," J. Appl. Physiol. 40:668-672 (1976).

54. Shvartz, E., and D. Benor. Heat Strain in Hot and Humid Environments. Aerospace Med. 43:852-855 (1972).

55. Smith, D. J. "Protective Clothing and Thermal Stress," Ann. Occup. Hygiene 23:217-224 (1980).

56. Soule, R. G. and R. F. Goldman. "Energy Cost of Loads Carried on the Head, Hands or Feet," J. Appl. Physiol. 27:687-690 (1969).

57. Sprague, C. H., and D. M. Munson. "A Composite Ensemble Method for Estimating Thermal Insulating Values of Clothing," ASHRAE TRANSACTIONS 80:120-129 (1974).

58. Tanaka, M., G. R. Brisson and M. A. Volle. "Body Temperatures in Relation to Heart Rate for Workers Wearing Impermeable Clothing in Hot Environments," Am. Ind. Hyg. Assoc. J. 39:592-599 (1979).

59. Wyndham, C. H. "The Probability of Heat Stroke at Different Levels of Heat Stress. "International Symposium: Quantitative Prediction of Physiological and Psychological Effects of Thermal Environment on Man," Centre d'Etudes Bioclimatique, Strasbourg, France (1973).

60. Yaglou, C. P., and D. Minard. "Control of Heat Casualties at Military Training Centers," A.M.A. Archs. Ind. Hlth. 16:302-316 (1957).

61. World Health Organization Scientific Group. "Health Factors Involved in Working Under Conditions of Heat Stress." Who Technical Report 412, 1969.
62. Brief, R.S. and R.G. Confer. "Companion of Heat Stress Indices." Am. Ind. Hyg. Assoc. J. 32:11-16, 1971.
63. Criteria for a Recommended Standard--Occupational Exposure to Hot Environments, USDHEW (NIOSH) HSM 72-10269, 1972.
64. Planned Update of Reference 63, scheduled for 1984.
65. DHHS (NIOSH) Publication No. 80-132, Hot Environments, 1980.

11

CONTAMINATION REDUCTION/REMOVAL METHODS

John M. Lippitt, RS
SCS Engineers
211 Grandview Drive
Covington, KY 41017
Timothy G. Prothero, BA
Hazardous Chemicals Specialists
4000 Murray Avenue
Pittsburgh, PA 15217
Lynn P. Wallace, Ph.D, PE
Brigham Young University

CONTAMINATION REDUCTION/REMOVAL METHODS

Chemical contaminants are removed by a combination of physical and chemical means; the chemical reactions used are generally designed to improve the efficiency of the physical methods of removal. Although the physical processes, even augmented by chemical means, are excellent for surface contamination, deep permeation of a substance by a contaminant usually requires the disposal of that equipment. The only way to remove something which has permeated into another is to create a stronger driving force towards the surface of the material than the driving force of the original concentration gradient.

Biological contaminants are also removed by a combination of physical and chemical methods. The physical removal of gross contamination and debris is necessary to reduce interference of other materials, such as soil, with the methods used for disinfection/sterilization of infectious agents. Table I is a list of common decontamination methods for use on equipment, tools, vehicles, and/or personal protective clothing and equipment, depending on limiting characteristics of the methods.

Table I. Common Decontamination Methods

---

1. Contaminant Removal

   * Scrubbing/Scraping with brushes, scrapers, sponges, etc. (commonly used in combination with solvent cleaning solutions)
   * Water rinse (pressurized or gravity flow)
   * Pressurized air jets
   * Steam jets (commonly used with solvent cleaning solutions)
   * Evaporation/vaporization
   * Chemical leaching and extraction

2. Detoxification

   * Oxidation/reduction
   * Neutralization
   * Thermal degradation
   * Halogen stripping
   * Irradiation

3. Removal of Contaminated Surfaces and Materials

   * Sand blasting
   * Disposal of permeated materials (e.g., seats, floor mats, clothing, etc.)
   * Disposable protective coverings/coatings (installed prior to activities to prevent or minimize contamination of uncontaminated surfaces or more extensive contamination of already contaminated surfaces)

4. Disinfection/Sterilization (infectious wastes)

   * Steam sterilization
   * Dry heat sterilization
   * Gas/vapor sterilization
   * Irradiation
   * Chemical disinfection

In many cases, gross contamination can be removed by physical means involving dislodging or displacement, rinsing, blowing or wiping-off, and evaporation. Often, the use of a disposable outer garment, such as a Tyvec suit, can be used to keep gross contamination from reaching protective clothing and thus reduce the required decontamination of such protective clothing. Contaminants which can be removed by physical means can be categorized as follows:

1. Loose Materials (Contaminants)

    Dusts and vapors can cling to equipment and workers through electrostatic forces or be trapped in small openings such as the weave of the fabrics of clothing. The slight forces of such bonding are fairly simple to overcome with the pressure of compressed air or with a liquid rinse. Water has excellent properties for eliminating static cling.

    The dislodging of electrostatically attached materials can be improved by preparation before site entry. Coating the clothing or equipment with anti-static solutions is usually done with commercially available wash additives or anti-static sprays.

2. Adhering Materials

    Some contaminants will have their own forces of adhesion other than electrostatic attraction. The adhesive qualities of contaminants will vary greatly with the specific contaminants and the ambient conditions (e.g., temperature). For example, properties of cohesion (i.e, molecular attractions between molecules of the contaminant) tend to increase with decreasing temperatures. In contrast, properties of adhesion (i.e., molecular attractions between contaminant molecules and molecules of contaminated materials) tend to increase as temperatures increase. Some materials, such as elemental mercury, will have much greater properties of cohesion than adhesion, and therefore present little resistance to physical removal methods. Other materials, such as glues, adhesives, cements, resins, and muds, have much greater adhesive properties than cohesive, and consequently are much more difficult to remove by physical means.

The usual physical removal means for gross
contaminants include scraping, brushing and wiping;
these methods work well at removing gross
contamination when the properties of the
contaminants are more cohesive than adhesive. With
wastes that are more adhesive than cohesive, the
adhesion/cohesion properties can be adjusted
favorably through certain methods such as
solidifying, freezing (e.g., using dry ice or ice
water), adsorption or absorption (e.g., with
powdered lime or kitty litter), or melting
(changing high viscosity solids to lower viscosity
liquids).

3. Volatile Liquids

The vaporization of volatile liquids is another
available technique. After evaporation or drying,
the residues can be removed with rinsing or
blowing. Some adhesives and muds are easier to
scrape off once they have been dried. To
facilitate removal, steam jets can be used to
vaporize those chemical contaminants which are
normally liquids at ambient temperatures. The use
of steam jets may also help with disinfection of
equipment.

4. Entrapped or Permeated Wastes

To physically remove a contaminant which is
entrapped in the weave of a fabric, or has
permeated into the molecular structure of a
material, an opposing driving force must be applied
in the opposite direction of permeation or
penetration. Sometimes creating a concentration
gradient in the desired direction is sufficient,
such as with washing and rinsing; in other cases, a
stronger driving force is needed. For example,
wringing out the contaminated cloth may compress
the space occupied by the contaminant, resulting in
a stronger driving force. Forced air can also
serve as a driving force by creating a pressure
gradient opposite to the original concentration
gradient of the contaminant. Protective cover
materials and clothing are often discarded due to
the difficulty of removing permeated contaminants.

Removal of gross contamination should be followed by a wash/rinse process which involves use of cleaning solutions to augment physical cleaning methods. The cleaning solutions normally utilize one or more of the following methods:

1. Solubilize

    Removal of chemical wastes from the surface can be achieved by dissolving the contaminants into a solvent for which the contaminants have a greater affinity than for the surface which is being cleaned. Common solvent types are water, dilute acids and bases, and organic solvents. Care must be taken to chose solvents which are chemically compatible with the waste being dissolved and the equipment being cleaned. That is especially important for personal protective clothing because it is made of organic materials which could be dissolved or damaged by the use of organic solvents.

    Organic solvents include alcohols, ethers, ketones, aromatics such as toluene, and straight-chain alkanes such as hexane. Often, common petroleum products such as fuel oil and kerosene are used. One must exercise caution in the selection and use of any organic solvent which may be flammable or potentially toxic. Some organic solvents selected for equipment and vehicles decontamination may not be suitable for protective clothing decontamination due to their potential toxicity and/or degradation/permeation materials. Organic solvents which are potentially toxic will require evaluation, selection and use of appropriate personal protective clothing and equipment to protect decontamination personnel using those solvents. The user of organic solvents must also be aware of the potential hazards for fire and explosion and take special precautions in the storage and handling of new and spent solvents. The spent organic solvents also have special disposal requirements.

    Halogenated products, such as dry-cleaning solutions (e.g., carbon tetrachloride) could be used to circumvent the flammability disadvantages of organic solvents; however, two other disadvantages remain: (1) compatibility with the personal protective equipment, and (2) the toxicity of the compound. The potential

exposure of the workers to the polyhalogenated product
would be self-defeating for the purposes of
decontamination of personal protective clothing and
equipment.

2.   Surfactants (Rosen, 1978)

Surfactants augment the physical cleaning methods by
reducing the adhesion forces between the contaminant
and the surface being cleaned; surfactants also prevent
the redepositing of the contaminants onto that
surface. Household detergents are among the most
common anionic surfactants, which are dependent in
large measure on the lack of dissolved cations in the
solvent medium (soft water). There are also detergents
which are used with organic solvents; dry cleaning
establishments use them to improve the dissolving and
dispersal of stains into the carbon tetrachloride
solution.

3.   Solidification

The cleaning of chemically contaminated surfaces can be
improved by solidifying the contaminating liquid or gel
so it can be physically removed (e.g., scraped off).
Solids have greater cohesive and lesser adhesive
properties than do liquids, gels, or pastes. The
mechanisms of solidification involve (1) removal of
moisture, (2) chemical reactions, and (3) cooling to
the freezing point. Some of the specific methods
include the use of an absorbent (such as grounded
clay), powdered lime, ice water, polymerization
catalysts, chemical reagents, and evaporation (drying).

4.   Rinsing

Progressive rinsing with clean solutions improves the
reduction of contamination over a single rinse of the
same volume of solution. Multiple rinsing is not quite
as efficient as showering. The removal of contaminants
through rinsing uses principles of dilution and
solubility. Their effectiveness is dependent on the
interfacial contact area and time between the
contaminated surface and rinse solution. By using
multiple rinses, more contamination can be removed than
with one single rinse; however, the continuous rinsing
or showering will remove still more contaminants, both
through solubility and physical action.

Disinfection and Sterilization of Infectious Agents

Unless special engineering controls can be provided, the most practical means of disinfection/sterilization will involve the use of chemical disinfectants (Tables II and III). Standard sterilization techniques would be impractical for larger equipment and not applicable for use on personal protective clothing and equipment while in use. Ultra-violet irradiation may be applicable, but must be evaluated relative to increased worker health risks. The use of pressurized steam jets may provide some assistance, but will need to be evaluated based on the temperature and time necessary to destroy the infectious agents. Steam jets are not equivalent to the time, temperature, and pressure controlled conditions of standard steam sterilization methods.

Selection of Appropriate Decontamination Solutions

In choosing the appropriate wash and rinsing solutions for a decontamination project, one must take into account the following factors:

1. Solubility behavior of contaminant
2. Compatibility of choice solutions with contaminant and equipment
3. Accessibility of solutions
4. Effectiveness of solutions and methods
5. Storage, handling, and disposal requirements of solutions
6. Hazards associated with cleaning solutions (i.e., flammability and toxicity)

In general, the more common solvents and the compounds they work best on are presented below :

1. Water:
   Dissolves low chain hydrocarbons, inorganics, salts, some organic acids and other polar compounds.
2. Dilute Acid:
   Dissolves caustic (basic) compounds, amines and hydrazines.
3. Dilute Base:
   Dissolves acidic compounds, phenols, thiols, and some nitro and sulfonic compounds.

Table II. Factors Impacting Chemical Disinfection

---

1. The types of organism
2. The degree of contamination
3. The amount of proteinaceous materials present in the waste
4. The type of chemical
5. The concentration and quantity of chemical disinfectant
6. The contact time
7. Possible interferences from other chemicals in wastes
8. The temperature of the item(s) being disinfected

Table III. Activity Levels of Selected Classes
of Liquid Disinfectants

Class	Use-Concentration	Activity Level[a]
Glutaraldehyde, aqueous	2%	High
Formaldehyde + alcohol	8% + 70%	High
Formaldehyde, aqueous*	3% to 8%	High to Intermediate
Iodine + alcohol	0.5% + 70%	Intermediate
Alcohols	70% to 90%	Intermediate
Chlorine compounds	500 to 5000 ppm[b]	Intermediate
Phenolic compounds	1% to 3%[c]	Intermediate
Iodine, aqueous	1%	Intermediate
Iodophors	75 to 50 ppm[d]	Intermediate to Low
Quaternary ammonium compounds	1:750 to 1:500[e]	Low
Hexachlorophene	1%	Low
Mercurical compounds**	1:1000 to 1:500[e]	Low

Courtesy of American Sterilizer Company, Erie, PA.
[a] Degree of disinfecting activity.
[b] Available chlorine.
[c] Dilution of Concentrate containing 5% to 10% phenolics.
[d] Available iodine.
[e] In appropriate diluent.
* See Section 4.7 of source reference for discussion of formaldehyde toxicity and necessary precautions for personnel protection.
** Should not be released into the environment, and therefore, is no longer used.

Source: U.S. EPA Office of Solid Waste, 1982.

4. Organic Solvents:
   Dissolve non-polar compounds such as other organics;
   can also dissolve fabrics of protective clothing.

The choice of solvent wash and rinse solutions will depend
largely on its compatibility with the material of the
equipment which is being decontaminated. For example, most
of the fabrics of protective clothing are made of polymer
organics which can be dissolved or destroyed by organic
solvents. The metals and gaskets of tools and equipment can
be damaged by overly acidic or caustic compounds. Thus, it
is very important to pretest cleaning solutions for
compatibility with the materials being cleaned.

Another important requirement is that the chemical waste
and the cleaning solutions be compatible. Figure 1 diagrams
the decision logic for selection of decontamination
wash/rinse solutions. Incompatible reactions resulting in
excessive heat, fire, or generation of toxic gases are not
desired in the contamination reduction areas where
decontamination crews might not be adequately protected.

The ready access to wash and rinse solutions forms the
most practical limitation. The availability of water has
resulted in its use in most decontamination cases,
regardless of the soluble nature of the chemical wastes.
Water and detergents are easily obtained in any site's
vicinity, and they are easily stored and handled. Organic
solvents, especially flammable ones, are less easily stored,
handled, and controlled.

The hazards involved with the solutions themselves must
be considered during any decontamination project. As
mentioned earlier, if organic solvents are used, the
flammability and toxicity hazards must be controlled. In
addition to the hazardous properties of the cleaning
solutions, however, one must remember that even water will
become hazardous after it has been used for cleaning
contaminated equipment.

The disposal of cleaning solutions will depend on the
type of solution (aqueous or organic), and the types and
amounts of contamination it contains after use. Depending
on the site and situation, cleaning solutions may be
collected after use and added to waste streams at the site
for the same disposal methods to be used for the wastes
(e.g., solidification, landfilling, incineration, etc.).

Disinfection solutions for use following removal of
gross contamination will require consideration of several
factors as presented in Table II. Table III provides
general information of activity levels of selected classes
of liquid disinfectants. It should be recognized that an
activity level based on field conditions should be
established prior to extensive site work.

275

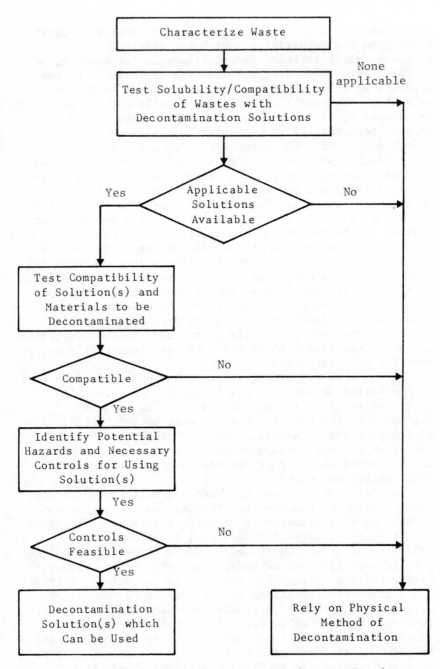

Figure 1. Decision logic for selecting decontamination wash/rinse solutions.

Measuring the Effectiveness of Decontamination Procedures

The effectiveness of decontamination procedures are often questioned because of the inability to perform real time measurements of their efficiency. Thus, it is often necessary for an arbitrary decision to be made concerning the endpoint of the procedures. The state-of-the-art in chemical analysis of surface contamination is still developing; however, several methods do exist (or can be created from available technology) for observing and measuring the decontamination procedures.

Methods available for measuring and inspecting effectiveness of decontamination includes:

1. Visual Inspection

   a. Natural Light. Visual inspection, using natural or artificial white light, entails the search for stains, discolorations, visible dirt, or alterations in the fabric of clothing as evidence of chemical contamination. The most obvious limitation is that not all contamination will result in visible staining or other similar traces. When such visual inspections are used, it is important that the searches include problem areas such as creases, boot treads, the crotch, etc., of personal protective equipment. Wheel wells, tool boxes, fixtures, etc., on heavy equipment also have problem areas that are difficult to inspect visually.

   b. Ultraviolet. Ultraviolet light is useful to detect certain contaminants which fluoresce, such as polycyclic aromatic hydrocarbons. These contaminants are common in many refined oils and solvent wastes. Ultra violet light can also be used to observe skin contamination, but one must be already aware of the areas of the subject's skin which naturally fluoresces. A disadvantage of such uses of ultra violet light is the added risk of increasing carcinogenic effects on the skin and the potential of damaging the eye.

2. Surface Analysis

   Instrumentation is currently being developed to detect, identify, and quantify contaminants on the surfaces of people, clothing, and equipment; however, at this time only large laboratory instruments are being marketed and most require the destruction of the sample. Fiberoptics is changing the technology of surface analysis and one can fabricate one's own field portable instrument from commercially available plans and parts.

3. Swab or Smear Samples

    Most available analytical methods for surface contamination are destructive of the sample; therefore, it is often recommended that if one wishes to preserve the clothing or equipment, that they should transfer the contaminant to another surface or solution. Wet samples are taken off of a selected area of the equipment or clothing and the samples may then be analyzed by wet chemistry field tests if the contaminant is known or sent to a laboratory for further qualification and/or quantification. A qualified analytical chemist should choose the sampling liquids and procedure so they are compatible with the contaminant, the surface sampled, and the test(s) to be performed.

4. Cleaning Solution Testing

    Another way of testing the overall decontamination procedure is to examine the amounts of contaminants left in the cleaning solutions; too much contaminant in the final rinse solution would suggest that additional cleaning and rinsing is advisable. When the type of contaminant is specifically known, a qualified chemist might devise some wet chemistry spot tests or other field tests to analyze the solutions. More complete analyses can be obtained by sending a sample of the solutions to be tested to a laboratory; however, time constraints for laboratory analysis can inhibit such a choice.

5. Disinfectant Solution Testing

    When dealing with infectious wastes, concentrations of active disinfectants can be measured in the spent solutions to determine if sufficient levels of active ingredients were available. The length of treatment with the measured level of activity can be compared to previous testing by qualified microbiologists to determine time/concentration ratios required to provide the necessary disinfection under actual or simulated conditions. As indicated in Table IV several factors can impact

Table IV. Factors Impacting Decontamination Design
Requirements

---

1. The chemical, physical, and toxicological properties of chemical wastes.

2. The pathogenicity of infectious wastes.

3. The amount and location of contamination.

4. The potential and location of exposure based on assigned duties, uses, activities, and functions.

5. The rates or potentials of wastes to permeate, degrade, or penetrate materials used for personal protective clothing and equipment, vehicles, tools, buildings, and structures.

6. The design and construction of personal protective clothing and equipment, vehicles, tools, buildings, and structures.

7. The proximity of incompatible wastes.

8. The purpose for leaving or removing equipment from the site.

9. The methods available for protection of workers during decontamination procedures.

10. The impact of the decontamination process and compounds on worker safety and health.

---

the disinfection efficiency. Therefore, it is advisable to confirm the assumed level of disinfection by laboratory culture of swab samples.

6. Microbial Swab Samples

Swab samples for infectious organisms taken from decontaminated surfaces and/or field controls (e.g., surfaces contaminated with another indicator organism, which requires similar levels of active disinfectants, length of treatment, temperatures, etc., or active cultures of the organisms of concern) should be transferred to testing laboratories for culturing under controlled temperature, atmosphere, nutrient, etc., conditions. Optimum growth and culture conditions are variable depending on the type of infectious organisms involved.

In-vivo (i.e., inside of the body) methods involving laboratory animals may also be used in conjunction with or instead of in-vitro (i.e., within an artificial environment) cultures. Exposed laboratory animals can be observed for clinical symptoms and/or tests can be run on samples of blood, tissues, etc., to determine if decontamination has achieved the desired level of disinfection.

Design of appropriate sampling and laboratory testing procedures should be developed by a trained microbiologist. Every effort should be made to design and select procedures which will provide the necessary confirmation as soon as possible. Unfortunately, while some procedures require only a few hours, others can require several days.

Decisions concerning decontamination endpoints are often based on the lack of visible contamination. Unfortunately, this does not address problems of permeation, thin layers of contamination, compounds which are not readily observable with the unaided eye, or infectious organisms which can only be observed under a microscope. As a precaution, unless sufficient field experience with laboratory confirmation is available for the compounds and conditions under which decontamination is being conducted, it is advisable to assume some level of contamination may remain. If the wastes involved are extremely hazardous, repetitive decontamination may be warranted even though obvious contamination has been removed. In addition, procedures for removal should be designed to prevent or minimize contact of unprotected skin surfaces with the exposed surfaces of clothing, equipment, tools, etc., which have been cleaned but may require further decontamination.

Respiratory protection should be maintained until potential sources and measured levels of hazardous vapors have been reduced to a predetermined safe level.

DECONTAMINATION PROGRAM DESIGN

Decontamination procedures provide four basic functions:

1. Minimizing worker contact with wastes during removal of personal protective clothing and equipment, and preventing exposures by cross-examination with wastes in clean areas where personal protective clothing and equipment is not worn.
2. Preventing dispersal of contaminants that have accumulated on personnel, clothing, equipment, and vehicles working or being used in the contaminated areas into surrounding areas off-site.
3. Preventing inadvertent mixing of wastes with other potentially incompatible wastes or compounds.
4. Removal or detoxification of wastes from equipment, vehicles, buildings, and structures in order to prevent continued exposures and enable other future uses after activities involving potential contact with contaminants are completed.

In principle, the design of a decontamination procedures is straightforward, but in practice it requires evaluation of many different variables. A general decision logic, as shown in Figure 2 can be used to organize decontamination procedures on a site. Within the decision logic, several factors must be considered based on the specific conditions of the site, characteristics of the waste, activities being conducted, etc. The decontamination of personal protective equipment and clothing, equipment, vehicles, structures, and buildings will frequently be different because of the differences in materials involved, the type of contamination, and the type of item being decontaminated. Table IV is a list of factors which impact on decontamination design requirements.

281

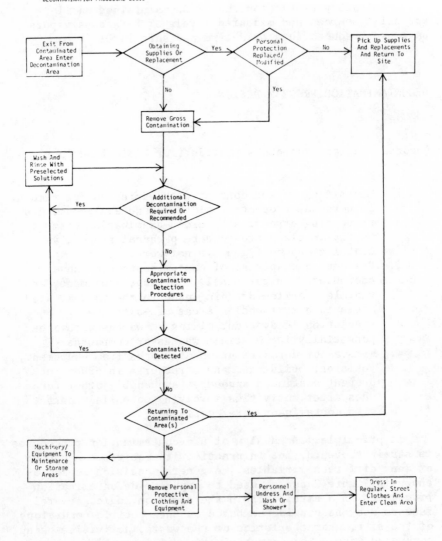

* Site clothes should be washed separately from
non-site and other family clothing.

Figure 2. Decontamination procedure logic.

282

Decontamination facilities on a hazardous waste site should be located in the contamination reduction area between areas contaminated by the wastes and clean support areas, as shown in Figure 3.

Decontamination designs and procedures must require the containment, collection, and disposal of contaminated solutions and residues generated during the process. Controls may include items such as spray booths with side walls or curtains to contain splashes and sprays, a collection tank for waste liquids, and drums for disposal of excessively contaminated materials and solid wastes from the process. Separate facilities should be provided for decontamination of large equipment to prevent cross-contamination of personnel decontamination facilities. Each stage of decontamination (e.g., gross decontamination and repetitive wash/rinse cycles) should be conducted separately, either by using different stations or sequential time periods.

Stations used should be physically separated to prevent cross-contact between stations. The stations should be arranged in order of decreasing level of contamination, preferably in a straight line. Separate flow patterns and stations should be provided when it is necessary to isolate workers from different contamination zones containing incompatible waste. Entry and exit points should be well marked and controlled. The decontamination area should be separate from the entry path to the contaminated area (exclusion zone) from the clean area (support zone). Dressing stations for entry should be separate from re-dressing areas for exit.

Procedures established should include a minimum decontamination procedure for use of restroom facilities. Entry into clean areas of the decontamination facility such as the locker rooms should require full decontamination. Restroom facilities should be located within an intermediate area as appropriate for the decontamination requirements established by a qualified safety and health professional.

Table V is a list of recommended supplies for decontamination of personnel, clothing, and equipment. Table VI is a list of recommended supplies for large equipment and vehicle decontamination. These lists are not inclusive for all needs, but should give general guidance on the types of supplies to be provided.

To augment and facilitate decontamination procedures, various standard operating safety procedures should be implemented. For example:

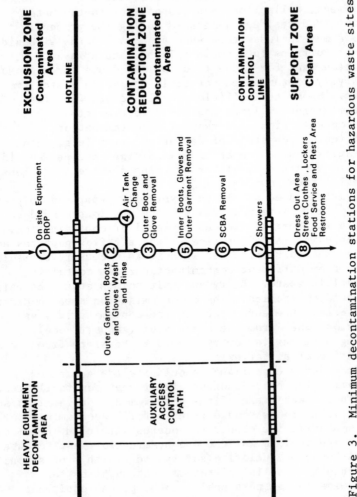

Figure 3. Minimum decontamination stations for hazardous waste sites that require respiratory and skin protection.

284

Table V. Recommended Supplies for Decontamination of
Personnel, Clothing, and Equipment

---

* Drop cloth(s) (plastic or other suitable material) for
  heavily contaminated equipment and skin outer
  protective clothing such as overboots, second pair of
  gloves, monitoring equipment, drum wrenches, etc.
* Disposal collection container(s) (drums or suitable lined
  trash cans) for disposable clothing and heavily
  contaminated personal protective clothing or equipment
  to be discarded.
* Lined box with absorbents for wiping or rinsing off gross
  contaminants and liquid contaminants.
* Wash tub(s) of sufficient size to enable workers to place
  booted foot in and wash off contaminants. (Without
  drain or with drain connected to collection tank or
  appropriate treatment system.)
* Rinse tub(s) of sufficient size to enable workers to place
  booted foot in and hold the solution used to rinse the
  wash solutions and contaminants after washing (without
  drain or with drain connected to collection tank or
  appropriate treatment system).
* Wash solutions selected to wash off and reduce the hazards
  associated with the contaminated wash and rinse
  solutions.
* Rinse solution to remove contaminants and contaminated
  wash solutions.
* Long-handled, soft-bristled brushes to help wash and rinse
  off contaminants.
* Lockers and cabinets for storage of decontaminated
  clothing and equipment.
* Storage containers for contaminated wash and rinse
  solutions.
* Plastic sheeting, sealed pads with drains, or other
  appropriate method for containing and collecting
  contaminated wash and rinse water spilled during
  decontamination.
* Shower facilities for full body wash or, at a minimum,
  personal wash sinks (with drains connected to
  collection tank or appropriate treatment system).
* Soap or wash solution, wash cloths, and towels.
* Clean clothing and personal item storage lockers and/or
  closets.

Table VI. Recommended Supplies for Large Equipment
and Vehicle Decontamination

---

* Containers for gross contamination involving removal of
    wastes and contaminated soils caught in tires, and the
    underside of vehicles or equipment.
* Pads for collection of contaminated wash and rinse
    solutions with drains (or pumps) connected to storage
    tanks or approved treatment system.
* Shovels, rods, and long handled brushes for dislodging and
    cleaning out wastes and contaminated soils caught in
    tires, and the underside of vehicles or equipment.
* Pressurized sprayer(s) for washing and rinsing
    (particularly hard-to-reach areas).
* Spray booths, curtains, or enclosures to contain splashes
    from pressurized sprays used to dislodge materials and
    clean hard-to-reach areas.
* Long-handled brushes for general cleaning of exterior.
* Wash solutions selected to remove and reduce the hazards
    associated with the contaminated wash and rinse
    solutions.
* Rinse solution to remove contaminants and contaminated
    wash solutions.
* Wash and rinse buckets for use in decontamination of
    operator areas inside the vehicle and equipment.
* Brooms and brushes for cleaning operator areas inside the
    vehicles and equipment.
* Containers for storage and/or disposal of contaminated
    rinse and wash solutions damaged or heavily
    contaminated parts and equipment to be discarded.

---

1. Training and site safety planning will enable workers to anticipate and avoid or minimize contamination of their protective clothing and equipment, tools, large equipment, etc. Disposable covers and outer garments can be used when large accumulations of wastes or contaminated media, such as mud, will be expected. This would simplify gross decontamination by allowing disposal of the heavily contaminated coverings and limit the degree and length of direct contact between wastes and the surface being decontaminated.

2. Procedures can be established for dressing and undressing when using personal protective clothing and equipment. Use of well designed procedures will afford greater protection from exposure by minimizing contamination of inside linings in direct contact with the skin.
Proper procedures for dressing will minimize potential for contaminants to by-pass the protective clothing and equipment barriers. In general, all fasteners should be used (i.e., zippers fully closed, all buttons used, all snaps closed, etc.). Gloves and boots should be under the cuffs of arms and legs of the outer clothing, and hoods (if not attached) should be outside of the collar. This prevents contaminants splashed or settling on the outside of the clothing from running inside the gloves, boots, and jackets (or suits if one-piece construction). Helmets, goggles, and respirators should be worn in a manner that does not interfere with their functions.
Removal of protective clothing and equipment should follow procedures which require removal of items based on the greatest potential for contamination (e.g., outer boots and outer gloves) and minimize contamination of inside linings (e.g., removal of jackets and pants after outer gloves and boots). Protective clothing and equipment should not be removed until potential sources of exposure are removed. For example, soils being removed from acid suits and releasing entrained solvent vapors could necessitate the use of respiratory protection until vapors are controlled and/or soils removed.

Actual decontamination facility design and setup will vary depending on:

1. The availability of utility service
2. Mobilization time and duration of site activities

3. The level of on-site activity anticipated and site conditions
4. The volume and level of decontamination required
5. Available space in uncontaminated area contiguous with the contaminated areas of a site
4. The volume and level of decontamination required
5. Available space in uncontaminated areas contiguous with the contaminated areas of a site
6. Potential hazards associated with the decontamination procedures and wastes generated.

The availability of utility services will determine the requirements for providing and storing portable water supplies for use in decontamination. If electrical service is necessary, portable generators could be required. Availability and access to waste water and sewage disposal systems may impact storage and handling requirements for decontamination wastes. To protect workers, restroom facilities should be provided in appropriate areas within the contamination reduction areas to enable and promote adequate decontamination prior to using restroom facilities. The requirement for the storage and handling of these human wastes will also be effected by availability of sewage systems.

Mobilization- and demobilization-time requirements influence the design of decontamination facilities. Emergency response to accidents, spills, fires, or explosion may not allow sufficient time for more elaborate facilities. An increasingly common practice is the use of mobile decontamination facilities which are self-contained and fully-equipped for personnel decontamination. However, in situations where mobile facilities are not available, decontamination kits should be devised. Table VII is an example of a decontamination kit which could be used. The use of temporary facilities versus more permanent facilities (such as the installation of a full-sized decontamination trailer or construction of on-site buildings and facilities) will often be dependent on the duration of site activity anticipated.

The level of on-site activity will determine, to a large extent, the potential for contamination of workers. Decontamination for investigations involving limited contact with contaminants for purposes of sampling is usually less elaborate than decontamination of workers involved in handling and packing of wastes during site cleanup. Similarly, sites on which releases of wastes have resulted in extensive contamination will require more decontamination of site workers than a site on which wastes have been contained and adequately controlled to minimize contamination.

288

Table VII. Example of Personnel Decontamination Kit

---

* Five-gallon container(s) of potable water (for
   decontamination only
* Soft- and stiff-bristled brushes
* Detergent (solid or liquid)
* Plastic wading pool(s)
* Buckets or sprinkler cans for rinsing
* Paper towels or other disposable cleaning cloths
* Chemical-resistant container(s) (minimum five gallons for
   wash/rinse solutions
* Plastic garbage bags (for storage of equipment and
   disposal of solid/hazardous wastes.

The number and frequency of workers undergoing decontamination will impact the flow design, size, and number of stations used. Likewise, the number and frequency of vehicle and large equipment decontaminated will impact designs for those facilities.

One of the most limiting factors for decontamination facility design and setup is the availability of space in uncontaminated areas contiguous with contamination zones. Maintaining sufficient separations between stations may often require use of fewer stations. Also the flow design may be arranged in rows or serpetine-fashion as opposed to the preferred straight-line design. The design of many of the mobile trailer facilities in use often require such modifications from the straight-line design to make the most efficient use of available space. In confined areas, it is essential that procedures and practices minimize splashing, sprays, dusting, and aerosols which may cross-contaminate stations. In enclosed areas, such as mobile trailers, effective ventilation controls are also critical.

Concerns for cross-contamination and protection of decontamination facility workers may require special designs and controls if potential hazards during decontamination are significant. Ventilation hoods, spray booths, wet wells, chemical treatment tanks for wastes, and specialized storage containers are examples of specialized components that may be used to control dispersion of wastes during decontamination. Standard controls should prevent or minimize run-off, air diffusion/dispersion, and movement of wastes from each decontamination station.

Levels of Decontamination

The level and type of decontamination required on a site will vary based on differences in exposures and the type of item being decontaminated (i.e., personal protective clothing and equipment being worn by a worker as opposed to large equipment). Although design of specific decontamination programs and procedures will require detailed consideration of all the factors involved (see Table IV), the following general variations and discussions are provided for guidance:

1. Field Investigation of Sites

   a. Known-Versus-Unknown Conditions. The problem with identifying an unknown situation is that, without knowledge of the specific compounds involved, it is difficult to determine an appropriate analytical method. Whenever the compounds of probable or definite exposure are unidentified, one cannot quantify the amount of exposure because analytical standards are not available. In such cases, a sample of the chemical wastes should go to a laboratory for analysis and identification. Unfortunately, the analysis may not be complete until several weeks after exposure and contamination have taken place.

There is an advantage in knowing as much about a situation as possible. In most initial investigations, however, very little is known in the beginning. In those situations, any knowledge about the site conditions, no matter how little, is an advantage. When the waste storage areas are visibly above ground, it might be presumed that contaminated areas are within a certain arbitrary perimeter delineated by the waste containment areas and roadways. Certainly, such clues as ground stains, container labels, etc., are of great assistance in developing safety and decontamination plans. When the site is below the ground (when no visible clues exist on the surface), it is very difficult to create any perimeter of contaminated areas. There is much less initial available information from which to determine an appropriate decontamination technique.

b. Active vs. Inactive Site. Valuable information concerning the types of wastes, their properties, and the site's operating policies and practices is available if the site is still operated by the original operator during the investigation. The site's operating plan should include information on where chemicals are stored, reacted, used, plus how and why. This plan can be reviewed to determine the most likely areas of high risk for exposure and the route by which workers are most likely to be exposed. Decontamination procedures can be formulated for all conditions at the site, plus special testing procedures can be instigated for determining the appropriateness of the decontamination procedures.

For inactive sites, it is necessary to be sure that all the appropriate information gathering has been done prior to any on-site visit. Where no valuable and verifiable conclusions can be reached about the

specific locations of contaminated areas, then
generalized precautions and decontamination
procedures must be followed. It is always
important to verify suspicions, hypotheses, and
conclusions inferred from information about a
site. Another important aspect of the active
versus the inactive site is that there will be
people engaged in activities at the active site
which are not under the direct control of the
on-scene coordinator. Those uncontrolled
activities will make zone delineation difficult and
cross-contamination between zones can be expected.
The placement of the contamination-reduction zone
will be more difficult because the perimeter of
contamination may be prone to frequent changes.

c. Direct Handling vs. Indirect Contact with
Wastes. Direct contact with contaminants in their
concentrated forms, by handling or processing
wastes, results in the highest concentration
gradient which would be encountered at a site. The
workers who actually handle the wastes for
reprocessing face the highest risks of harmful
exposures. The high concentration gradients mean
that the protective equipment of those workers will
probably be permeated to differing degrees,
depending on the wastes and equipment involved, and
the workers will find that decontamination is
relatively more difficult than for others engaged
in tasks such as observing. It is important to
note that the routes of contamination for wastes
handling workers includes splashing, handling (on
gloves), stepping in wastes (on boots), as well as
high vapor concentrations when working in close
proximity to volatile wastes.

One must not forget that potentials for direct
contact still exists, even for the site observer
who does not intentionally handle any of the
wastes. Just by being present at the site, an
observer may be exposed to vapors, contaminated
dusts, wastes in the ground and in run-off waters,
etc. One of the chief differences in the
contamination is that a person who works with the
chemicals is likely to show visible staining on his
protective clothing, thus confirming the need for
aggressive decontamination. A person who is
exposed to only the contaminated dust and dirt,
however, might choose to forego aggressive
decontamination in the belief that any potential
exposure was minimal.

292

2. Decontamination While Conducting Site Cleanups

    a. Waste Handlers and Other Positions. The first
    step in personnel decontamination is the removal of
    any known, visible, and obvious contamination.
    This entails the cleaning of the contaminated
    surface until the presence of the contaminants is
    no longer detectable. Then the worker proceeds to
    take measures to remove suspected or undetectable
    contaminants, such as those that might arise on
    exposure to vapors. Finally, to further protect
    the worker, his family and friends, workers must
    undergo personal cleaning, including showering and
    changing to clean street clothes. Even site
    observers who do not directly handle the wastes
    must undergo some decontamination, usually starting
    with step two to remove unseen and unsuspected
    contaminants.

    b. Equipment Maintenance and Repair Crews. Due to
    the tendency for waste handling equipment to become
    heavily contaminated, there also exists a special
    need to protect the maintenance and repair crews.
    The maintenance workers may be working on vehicles
    and equipment which undergo only some gross
    contamination reduction instead of complete
    decontamination. For that reason, it is desirable
    to protect the maintenance people through the use
    of personal protective devices and complete
    personnel decontamination.

The site safety officer should determine the proper
type of worker protection for the maintenance and
repair crews and the type of necessary worker
decontamination. The direct exposure areas are
presumed to be mainly on the hands, forearms, feet,
and knees; the decontamination efforts should
concentrate on aggressive contaminant removal in
those areas, plus total body decontamination for
unsuspected contaminants. The selection of proper
cleaning solutions for the maintenance crews is
simple in theory, but difficult in practice. One
need only test the contaminants found on the
equipment to determine the best cleaning methods;
however, it is likely that the heavy equipment and
vehicles have been used throughout the site, thus
obtaining a wide spectrum of contamination, and
each member of the maintenance crews is likely to
work on several pieces of equipment during his
shift.

Equipment brought in for refueling should be refueled by someone in appropriate protection for the general on-site work since it is doubtful that the vehicle would be decontaminated for refueling purposes alone. One must remember that contaminated equipment should not be brought into the clean (support zone) areas. The refueling person, if not engaged in other work activities in the contamination zone, should then undergo the decontamination procedures set up for the site workers and equipment operators.

3. Heavy Equipment and Tools

   a. Heavy Equipment and Vehicles. Normal site operating requirements will probably find daily decontamination of the heavy equipment to be impractical. Equipment and vehicles can be decontaminated to differing degrees during site operations. From the previous discussion of permeation, one should realize that the contaminants will permeate into the equipment to some extent unless prompt decontamination takes place; however, since most of the construction material of the equipment and vehicles is metal, which is not readily permeated, only a partial daily decontamination procedure is necessary. On a daily basis, gross contaminants should be removed from all equipment surfaces to prevent build-up and caking; plus, all non-metallic surfaces (seats, tires, operator cabin, etc.) should be completely decontaminated daily to prevent permeation and to protect the operator.

When total decontamination of a piece of heavy equipment is to take place, it is likely that industrial cutting solvents will be used on the metal surfaces. The crew members must remember to follow appropriate precautions in the handling and use of potentially hazardous cleaning solvents. They must also remember to clean such areas as the undercarriage of the equipment, under the seat, in the engine area, and the vents and air ducts.

Again, it is important to collect and properly dispose of the cleaning solutions and the mud and sludges which are scraped off the equipment.

   b. Tools. All tools used to maintain and repair equipment, or to handle wastes on-site, must be

cleaned daily even though they may be made entirely
of metal. The small size of hand tools allows
easier cleaning than heavy equipment and common
tools are frequently handled without gloves or
regard for their previous use. For the general
protection of all crew members and to limit sources
of cross-contamination, all hand tools should be
cleaned after each use and before storage.

Decontamination of on-site tools serves to protect
the workers in two ways: (1) protection from
unnecessary exposure, and (2) protection from
interference of dexterity with tools due to the
build-up of slippery contaminants. Heavy rubber
grips on hand tools will enhance the workers'
dexterity while wearing layers of heavy gloves, but
it makes decontamination more difficult. The use
of removable, disposable grips will aid in the
overall safety involved with hand tools.

4.   Protection of Decontamination Workers

Decontamination workers are subject to potential
exposures from three sources, including:

   *Contaminants on workers, clothing, and
    equipment exiting from the contamination zone.

   *Potential hazards associated with the storage,
    handling, and use of decontamination solutions.

   *Used solutions and wastes generated during the
    decontamination process.

Personal protective clothing and equipment used by
decontamination workers should protect them from
potential exposures from all three sources.

Workers initially coming in contact with personnel
and equipment leaving the decontamination zone will
require more protection from waste contamination
than those at later stages after gross
decontamination. Workers conducting
decontamination procedures, such as using a steam
jet for removal of contamination, may need a
different type of respiratory protection because of
the high moisture levels. In some situations, used
solutions and wastes generated during
decontamination may result in the generation of
additional vapors, from which workers need to be

protected. The requirements for worker protection and selection of appropriate personal protective clothing and equipment should be reviewed and approved by qualified health and safety personnel assigned to the site.

Although the decontamination crew is near the clean zone and assisting in processing personnel and equipment into the clean areas, the crew is, nevertheless, in a contaminated area. As such, all decontamination personnel must, decontaminate themselves before exiting into the clean areas. The extent of their decontamination will be determined by the type of decontamination work in which they are engaged and the types of contaminants they may have contacted.

All spent and used cleaning solutions should be collected and properly disposed of. The solutions could be placed back onto the site for later disposal, or included with some compatible on-site wastes for proper disposal; other treatment and disposal options include solidification, chemical treatment, incineration, and carbon filtration. It is usually the responsibility of the on-scene coordinator to determine the disposal of the cleaning solutions.

Emergency Decontamination

In addition to normal decontamination procedures, emergency decontamination procedures should be established. In an emergency situation, decontamination may not occur at the site when immediate treatment is required to save a life. If decontamination can be provided without interfering with essential first aid and life-saving techniques, such as CPR, then it should be done. Clothing and equipment may be washed, rinsed, and/or cut-off when necessary. Otherwise, the individual should be covered with a blanket or other suitable material. Covering serves to prevent contamination of ambulance and medical personnel. Alternatively, to minimize possible heat stress to the patient, protective clothing and equipment may be used by the emergency response personnel (assuming appropriate training has been provided during planning stages of the project).

It is important to coordinate procedures for decontamination protection of medical personnel, and disposal of contaminated clothing and equipment. These procedures are necessary to minimize the risk of exposure to emergency medical personnel. Such procedures should be established during planning of site activities prior to any site work. Figure 4 outlines the decision logic which should be followed in an emergency.

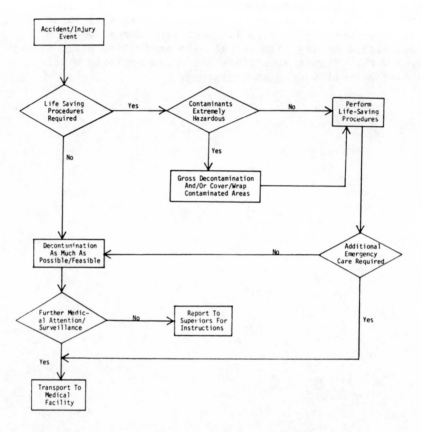

Figure 4. Emergency decontamination design logic.

REFERENCES

1. Advisory Committee for NIOSH Carcinogen Laboratory. Protocol for the NIOSH Carcinogen Laboratory. Unpublished. National Institute for Occupational Safety and Health, Cincinnati, Oh.

2. Department of Health, Education, and Welfare, Committee to Coordinate Toxicology and Related programs, Laboratory Chemical Carcinogen Safety Standards Subcommittee. Guidelines for the Laboratory Use of Chemical Substances Posing a Potential Occupational Carcinogenic Risk--Revised Draft. National Institute for Occupational Safety and Health, Cincinnati, OH, 1979.

3. International Agency for Research on Cancer (IARC). Handling Chemical Carcinogens in the Laboratory Problems of Safety. R. Montesano, H. Bartsch, E. Boyland, G. Dellaporta, L. Fischbein, R. A. Griesemer, A. B. Swan, L. Tomatis, and N. Davis, eds. IARC Scientific Publications No. 33, 1979.

4. Mayhew, Joseph I., G. M. Sodaro, and D. W. Carroll. A Hazardous Waste Site Management Plan. Chemical Manufacturers Assoc., Washington, D.C., 1982.

5. Mine Safety Appliances (MSA), Chemical Resistance Total-Encapsulating Suits. Data Sheet 13-00-07, Pittsburgh, PA.

6. Rosen, M. J. Surfactants and Interfacial Phenomena. Wiley-Interscience Publication, NY, 1978. 304 pp.

7. Rybak, Carl. Guidelines for Operation of HERL Carcinogenic Dilution Room. Unpublished Draft. U.S. Environmental Protection Agency Health Effects Research Laboratory (HERL), Cincinnati, OH, 1981.

8. Tucker, Samuel P. Deactivation of Hazardous Chemical Waste by Methods Other Than Conventional Incineration and Biological Degradation. Unpublished Draft. National Institute for Occupational Safety and Health, Cincinnati, OH, 1983.

9. U.S. Environmental Protection Agency/Hazardous Response Support Division (EPA/HRSD). Personnel Protection and Safety-Training Manual. National Training and Technology Center, U.S. Environmental Protection Agency, Cincinnati, OH, 1982.

10. U.S. Environmental Protection Agency/Office of Emergency and Remedial Response (EPA/OERR). Interim Standard Operating Safety Guides. Edison, NJ, September 1982.

11. U.S. Environmental Protection Agency, Office of Solid Waste and Emergency Response. Draft Manual for Infectious Waste Management. EPA-SW-957, U.S.

Environmental Protection Agency, Washington, D.C., 1982. 147 pp.

12. Vo-Dinh Tuan. Surface Detection of Contamination: Principles, Applications, and Recent Developments. Journal of Environmental Sciences. January/February 1983, pp. 40-43.

13. Vo-Dinh and Gammage. The Use of a Fiberscope Skin Contamination Monitor in the Workplace. Chemical Hazards in the Workplace, American Chemical Society, 1981a, pp. 269-281.

14. Vo-Dinh and Gammage. The Lightpipe Luminoscope for Monitoring Occupational Skin Contamination. American Industrial Hygiene Association Journal (42), 1981b, pp. 112-120.

15. Vogel. Vogel's Textbook of Practical Organic Chemistry. Longman Group Ltd., London, 1979, pp. 940-947.

16. Permeation of Protective Garment Material by Liquid Halogenated Ethanes and a Polychlorinated Biphenyl, 81-110, National Institute for Occupational Safety and Health, January 1981.

17. Lippitt, John M., T.G. Prothero, W.F. Martin, and L.P. Wallace. "An Overview of Worker Protection Methods," in the Proceedings of 1984 Hazardous Material Spills Conference, Nashville, TN, April 9-12, 1984.

12

TRAINING

J. Larry Payne, BS, MA
Clyde B. Strong, Jr., BS, MS
Texas Engineering Extension Service
Texas A&M University System
College Station, Texas 77843-8000

Hazardous materials and hazardous waste training have
long been topics of discussion among industrial personnel,
emergency response teams, regulatory agencies, and allied
groups. Although a number of successful programs have been
designed to meet specific needs, little has been done to
develop and clarify a sound educational philosophy that
might be applied in toto to hazardous material training.
While it is not possible to design one single curriculum to
meet all training needs, it should be equally obvious that a
number of generalities and guidelines exist that are useful
in developing hazardous material training programs.

HISTORICAL PERSPECTIVE

Prior to 1975, there existed few well-defined training
programs that focused on cleanup procedures for spills of
hazardous materials. In 1975 the Texas Engineering
Extension Service of The Texas A&M University System began
offering a five-day oil spill control course on a regular
basis. This program was unique in its hands-on approach to
practical spill training. Its major limitation was its
orientation to the petroleum audience; hazardous materials
and substances other than petroleum products were not
covered.

From the outset it became apparent that a hazardous
material control program was needed. By 1977 staff
personnel had developed a five-day hazardous material
control course similar in format to the existing oil spill
control program.

A CONCEPTUAL FRAMEWORK FOR PROGRAM DEVELOPMENT

In order to develop such a program in an orderly and systematic fashion, one must view the task in conceptual terms. Typically, program start-up involves a preparation stage, a development stage, an implementation stage, and an evaluation stage. If the program status goes beyond that of a pilot, an improvement feedback loop develops between the evaluation and implementation stages (see Figure 1).

THE PREPARATION STAGE

During this preliminary phase, a number of considerations must be addressed, including a complete needs assessment. In assessing the needs for a hazardous material control program, questionnaires, pre-tests, and standard compilations of sociometric data can be relied upon. During this phase of operations in developing the Texas A&M University program, staff members evaluated existing hazardous material training programs and seminars. A thorough examination revealed that many of these programs carried several inherent defects. In some instances the programs focused chiefly upon packaging, labeling, and shipping requirements, neglecting any mention of response or mitigation tactics. Other programs purporting to emphasize techniques for handling hazardous materials and wastes were too often limited to classroom lectures and slide shows describing previous incidents. In many cases they failed to provide either a viable educational rationale or sufficient practical training.

Audience analysis represents another often neglected part of this preparatory phase of development. The creation of a successful hazardous material control program must begin with an adequate audience analysis. Various tools may be used to compile data on potential audiences, but it remains the sole responsibility of the program development team to analyze, assimilate, and apply such data. The goals and objectives for a program on proper use and maintenance of a self-contained breathing apparatus will vary significantly depending upon whether the primary audience is composed of experienced response team members or novices. Similarly, there may be considerable gaps between the manipulative abilities of an experienced response person and those of a lab technician. Again, no single program can

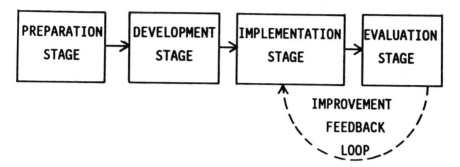

Figure 1. Stages of program development.

accommodate all of the needs of a highly diversified audience; however, in order to succeed, a program must be directed toward a specific target audience.

Before turning away from the topic of program preparation, the role of the advisory committee as a means of diagnosing needs and analyzing audiences needs to be mentioned. Experiences with several programs offered through the Texas Engineering Extension Service of The Texas A&M University System have demonstrated that well-selected advisory groups representing industry and other appropriate agencies are critical to the viability of spill and incident control training programs. One case in point is the Oil Spill Control Course Advisory Committee composed of members from the American Petroleum Institute (API) and major oil companies. Another example is the Hazardous Material Control Course Oversight Committee chaired by the representative from a major chemical company and made up of representatives from various sectors of the chemical, transportation, and petrochemical industries.

In both cases these advisors' groups carefully refrain from promoting The Texas A&M University System programs. Their purpose is not to endorse the programs, but to provide insights and information regarding how the programs might be better organized and improved in order to meet the most pressing needs of industry. Representatives from the industrial sectors offer excellent sources of information regarding basic training needs as well as valuable audience analysis data. By selecting representatives from a variety of industrial backgrounds and from different geographical locations, it is possible to avoid the parochialism that might otherwise develop in a state or regional training program.

THE DEVELOPMENT STAGE

Once the stage has been set, it is necessary to move into the developmental phase of operations. The bulk of the work to be done in this area entails the development of teaching/learning materials that will be compatible with the needs and audiences identified earlier.

Taba identifies two major areas of activity in this stage: (1) formulating objectives, and (2) selecting and organizing content [1]. As one might expect, the two areas involving objectives and content have a tendency to overlap. Therefore, it is impractical to regard them as totally separate and distinct; objectives will influence content and vice versa.

In formulating broad-based objectives for a hazardous material control program, the goals of a course should:

1. Meet the general needs of those concerned with hazardous material spills including waste handling reporting regulations, control techniques, and recovery operations;
2. Focus upon a variety of hazardous substances such that participants could become familiar with a wide range of hazardous materials.
3. Cover lower cognitive and theoretical material in the classroom, reinforcing and expanding bases with hands-on manipulative training;
4. Introduce participants to pragmatic aspects of personnel protection, toxicology, and site safety operations;
5. Familiarize attendees with fire control tactics and strategies that might be relevant and applicable to hazardous material incidents;
6. Offer an opportunity for participants to test and evaluate their capabilities by responding to simulated hazardous materials incidents; and
7. Serve as a clearinghouse through which attendees could obtain information on the latest equipment, apparatus, and procedures for controlling hazardous material incidents [2].

By using this basic format and making modifications where necessary, it is possible to establish suitable underpinnings not only for a single program, but for a number of spin-off activities as well.

Although these objectives can certainly provide an adequate nucleus for a training program for handling hazardous materials, most instructional staff members find it useful to fine tune such broad-based objectives into more specific behavioral objectives. Well-defined behavioral objectives should be written clearly in a manner that incorporates: (1) a readily observable behavior; (2) the conditions or restrictions under which such behavior is to be observed; and (3) the performance level expected of the learner [3]. For example, a behavioral objective involving hazardous waste site assessment might be phrased as follows: "The learner, while wearing an encapsulated suit, will use a standard field identification kit to correctly identify an unknown waste sample taken from an overpack drum within one hour." The components of the behavioral objective are identified below:

<u>OBSERVABLE BEHAVIOR</u>	– will identify unknown waste sample
<u>CONDITIONS OF BEHAVIOR</u>	– while wearing an encapsulated suit and using a standard field I.D. kit
<u>PERFORMANCE LEVEL</u>	– <u>correctly</u> identify within one hour

Such behavioral objectives are useful in providing coherence and unity. Yet, if not used carefully, they may become awkward, restricting program flexibility. For instance, if such detailed objectives actually become an integral part of the course manual, then in order to remain truly accurate each time an objective is altered, the manual itself will have to be modified. For this reason, it is often helpful to include the behavioral objectives on the instructor's lesson plans where they might be altered without disrupting overall program flexibility.

The second major activity in the development stage involves selecting and organizing content. Content selection is normally thought of as a straightforward, linear process. It is commonly thought that one would be able to simply list the major topics of concern, and merely subdivide them in order to prepare a curriculum outline. Unfortunately, this approach presupposes unlimited time and a learning process that proceeds in a predictable, clockwise fashion.

Perhaps one of the greatest difficulties faced by an individual attempting to select content for a training program of this nature is in determining what to include and what not to include. As suggested earlier, a careful audience analysis and well-formulated objectives make this task easier. Still, program developers often find themselves with too much material and too little time. Content must be selected such that the material can be presented to the <u>target audience</u> within the specified time constraints. The key here lies within the ability to reach the specific target audience identified earlier. Participants on the periphery of this audience should not be ignored; nor can they expect the program to accommodate them at the expense of the target group. As a matter of fact, such peripheral participants will generally derive at least some benefit from the content selected for the target audience.

Priorities must be established so that the most important information receives primary consideration. Figure 2 illustrates the relationship between a topic and its four basic levels of content with Level I information holding the highest priority for inclusion and Level IV the lowest. One might expect most content selection to be made

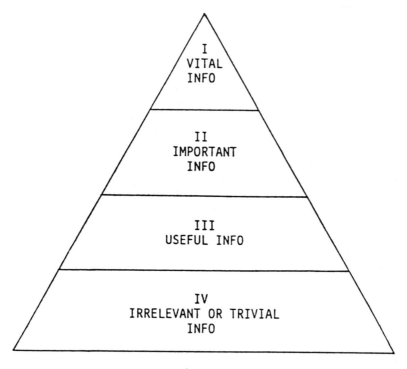

Figure 2. Topical information levels.

from Levels I, II, and III in that order. In spite of the
practical simplicity of this approach, however, the amount
of Level IV information that creeps into an instructor's
presentation as misguided attempts at levity, war stories,
and attention-getting devices is appalling.

Coincidental to content selection is content
organization. Taba suggests that topics should generally be
structured so that there is movement from the known to the
unknown, from the simple to the complex, from the easy to
the difficult [1]. It is often advantageous to identify
certain core ideas that seem to represent the very essence
of certain topics. Without fail these ideas and concepts
fall within the realm of Level I information in that they
are both basic and vital. Surrounding these vital kernels
are myriads of information which are mostly represented in
Levels II and III. What must be done, then, is to select
certain clusters of information from these areas that will
selectively reinforce and expand the core ideas. This
concept of organizing content with respect to core and
cluster areas is not new, yet it is seldom applied in the
areas of training for hazardous material and hazardous waste
handling.

THE IMPLEMENTATION STAGE

The third major phase of program development, that of
implementation, involves translating what has previously
existed as theory into practice. Of particular interest are
the subjects of selecting and organizing learning
experiences, and applying teaching methodologies.

Hazardous material control training offers an excellent
opportunity to apply a well-balanced mixture of classroom
and field training techniques. This basic teaching-learning
design first emerged out of the Texas Engineering Extension
Service's Oil Spill Control Course in 1976 as an attempt to
successfully mesh traditional classroom presentations with
pragmatic hands-on training. In order to achieve the
necessary balance between classroom and field training for
this program, it was necessary to provide traditional
classroom theory as a complement to pragmatic field
application. In order to subordinate formal classroom
sessions to hands-on training, instructors introduced each
major topic through a brief classroom presentation and
relied upon field exercises for reinforcement and practical
clarification. Instructors also attempted to enhance
student participation through discussion groups,
problem-solving exercises, and questions and answer

308

periods. Subsequent evaluation of this early program revealed four primary benefits:

1. By encouraging participation, students benefit from active as opposed to passive learning;
2. Similarly, student participation in a seminar atmosphere provides for valuable exchange of information among the students themselves;
3. Introducing basics in class and reinforcing and expanding them with hands-on training tempers textbook learning with actual experience;
4. Finally, this type of arrangement allows the course to be taught effectively, by following the "spiral step" method [4].

The spiral step methodology is the underlying strategy inherent in this type of program. This approach exposes course participants to units of subject matter in increasing order of complexity, continuously reinforcing them with appropriate skill developing exercises. Hence, Taba's order of moving from the simple to the complex, the known to the unknown, and the easy to the difficult is maintained. Moreover, this pedagogical approach helps to preserve the unity of the curriculum as established during the preparation and development stages by providing certain "threads" or common denominators that link basic concepts in an overall spiral configuration. One of the greatest arguments in favor of this strategy lies not in its ability to preserve order and unity, but in its tendency to promote what has been called cumulative learning. Taba suggests that such learning involves the cumulative development of mental skills such that each succeeding idea or question requires an increasingly difficult mental operation [1]. When training in hazardous material control, however, this cumulative effect involves not only the cognitive but also the manipulative domain. Participants are thereby able to simultaneously expand both dimensions of their abilities. The schematic in Figure 3 depicts how one isolated area--toxicological considerations--is introduced, expanded, and reinforced in a spiral step manner. It is important to note that the curriculum is designed such that each successive upward spiral represents an increase in both the complexity of the material covered and the difficulty with which physical skills are mastered, thus helping to foster cumulative learning.

While the spiral step approach is a crucial element in curriculum design and integration, there are other teaching methodologies and precepts that also bear mentioning. One such element involves the depth with which field training is carried out. Care must be taken to avoid structuring

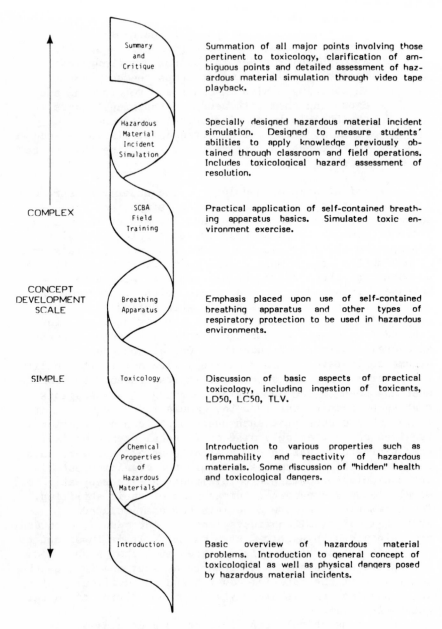

Summary and Critique		Summation of all major points involving those pertinent to toxicology, clarification of ambiguous points and detailed assessment of hazardous material simulation through video tape playback.
Hazardous Material Incident Simulation		Specially designed hazardous material incident simulation. Designed to measure students' abilities to apply knowledge previously obtained through classroom and field operations. Includes toxicological hazard assessment of resolution.
COMPLEX	SCBA Field Training	Practical application of self-contained breathing apparatus basics. Simulated toxic environment exercise.
CONCEPT DEVELOPMENT SCALE	Breathing Apparatus	Emphasis placed upon use of self-contained breathing apparatus and other types of respiratory protection to be used in hazardous environments.
SIMPLE	Toxicology	Discussion of basic aspects of practical toxicology, including ingestion of toxicants, LD50, LC50, TLV.
	Chemical Properties of Hazardous Materials	Introduction to various properties such as flammability and reactivity of hazardous materials. Some discussion of "hidden" health and toxicological dangers.
	Introduction	Basic overview of hazardous material problems. Introduction to general concept of toxicological as well as physical dangers posed by hazardous material incidents.

Figure 3.   Spiral step curriculum design.

practical exercises as demonstrations. While certain demonstrations are useful, they simply cannot replace active student participation. In addition, field training must be carried out under realistic conditions in order to be most effective (see Figure 4). Of course such realistic field training necessitates careful supervision by qualified instructors as well as enthusiastic participation by course attendees. Often overlooked in many programs is the need to keep such practical exercises small in order to maintain an adequate instructor-participant ratio. Depending upon the exact nature of the exercise, an adequate ratio for field training may be maintained by providing one instructor for every five to 10 participants. Going beyond a limit of 10 participants per instructor may create logistics problems as well as safety hazards under certain conditions. Likewise, it may force the instructor to rely more heavily upon strategies normally reserved for demonstrations, diminishing the desired effect of the field exercises.

In addition to the spiral step approach, techniques to help bridge the gap between theory and application need to be applied to training programs involving hazardous materials control. It has been previously noted that a well-designed spiral step curriculum can provide a sound basis for cumulative learning. Well-formulated objectives, properly selected content, and carefully chosen teaching methods all combine to form an integrated and balanced program. Basic topics are presented in classroom sessions, and expanded and specifically applied under field conditions. Thus, a student may learn a considerable amount about SCBA's in the classroom and may develop a fair amount of expertise in using them under field conditions. The same might be said for a number of other topics such as protective clothing, waste recovery, and spill control. However, there must exist a means of helping the course participants to assimilate this knowledge into a broad perspective. In other words, they must be able to take this information and apply it to a greater scenario such as a hazardous waste dump site, a train derailment, or a hazardous material spill.

One way of bridging the gap between the acquisition and the application of information is through a written problem session. Such a problem session must necessarily come after participants have covered the bulk of the course material. Inserted into the programming at this point, the problem session becomes a tool with which the instructors and participants might generate a positive attitude and response. Attempting to place the problem session too early in the programming before all basic materials have been covered would lead to confusion and probably a feeling of negativism.

311

Figure 4.   Field training pays off as students stabilize
a leaking tank car.

Such a session usually works best if it is carefully thought out and uses accurate maps and descriptions of the problem scenario. All variables such as date, location, meteorological conditions, logistics, and available supplies and manpower are generally given by the instructor. All that remains, then, is for the participants to respond to the scenario within these guidelines. Of course some double-sided problems are included, not to trick participants, but to alert them to hidden difficulties that may not be readily apparent. As a group response this strategy works well; not only do the group members use what they have learned as a result of the training program, but they also draw upon their own experiences and those of their peers. Most importantly, they are allowed the luxury of seeing a complete scenario unfold and develop before their eyes. Quite literally, they may respond to a full-blown incident without leaving their chairs. At this point they begin to make the transition from simply acquiring the information to applying it.

Finally, some means of going beyond this armchair quarterback situation is desirable as both a measuring device and confidence builder. The full-scale field simulation provides this means. Incorporated into a program after all basic knowledge and skills have been introduced and expanded, the field simulation is designed to approximate as nearly as possible "the real thing". Because it is designed as a test of sorts, the simulation should not be used by instructors as a device to measure individual skills and competencies. Rather, it should be seen as a measure of the entire group's ability to function under the stress of an actual incident. Individuals must assess the overall situation, sort through the relevant variables, determine appropriate response measures, and initiate them accordingly. The more realistic the controlled conditions are, the more successful the simulation is likely to be. Performing under such conditions has a tendency to dispel any attitude of "it's only a game". Realistic conditions also have a way of producing stress in individuals to a point where they perform at the upper limits of their capabilities. Furthermore, such simulations often serve as confidence builders (see Figure 5). For maximum effectiveness, it is helpful to videotape the simulated incident for prompt playback. The videotape playback serves as an informal critique of the group's overall response. Instructors must be careful to avoid playing too obvious a role in the critique. Emphasis must be placed upon constructive criticism. It is imperative that the instructor moderate this critique work skillfully to allow the participants to provide the greatest bulk of the feedback. Experience has shown in many cases that the

313

Figure 5.   Students find themselves tested by a
realistic simulation.

enthusiasm from a well-designed simulation carries over so that those involved actually do a better job of evaluating their performances than the instructors.

THE EVALUATION STAGE

One of the most important aspects of any educational program involves an assessment of the success of the program as perceived by the instructor and participant. In most traditional settings the primary evaluative tool directed toward the student is the formal examination. In hazardous materials control training courses, the use of formal examinations is not recommended. Instead, in these courses and other short industrial courses, it appears that the use of individualized instruction tactics might be applied more effectively. Such tactics normally require that the instructors develop a rapport with course participants so that they might continuously receive and digest feedback from them. By so doing, the instructor can accommodate participants by clarifying and expanding upon course information as required. Although this tactic develops around and exists upon a relatively informal phase, it nonetheless offers a reasonable means of monitoring and improving student performance.

Another evaluative tool mentioned earlier is the hazardous material simulation. As discussed, the simulation provides the basis for a post-incident critique that may be helpful in evaluating the group's overall response; however, it rarely develops enough detail to allow for individual evaluation. Indiscriminate use of such critiques can often do more damage than good by attempting to place blame and highlight individual mistakes.

Although it is important to assess levels of student performance, it is perhaps even more important to determine the strengths and weaknesses of the program itself. This may be especially true in the case of short, continuing education courses. Course evaluations provide basic feedback essential to program improvement. Such an evaluation should be simple, easily tallied, and should allow for comments regarding each specific topic of presentation of a program. Moreover, an attempt should be made to ensure that students regard such evaluations as important tools. Accuracy and honesty must be emphasized in the name of constructive criticism. An important point here is that the actual presentation, not the instructor, should be evaluated. It is wise to prompt course participants to evaluate each session as it concludes rather than evaluating

all the sessions at the end of the course. It is also
important to solicit suggestions and other comments that
might be helpful in improving the course. Anonymity often
helps to ensure objective ratings. Appendix I illustrates
one typical course evaluation sheet used successfully by the
Texas Engineering Extension Service.

CULTIVATING THE IMPROVEMENT FEEDBACK LOOP

Course evaluations including written critiques and
informal comments must somehow be systematically analyzed if
they are to help establish a means of improving the
program. Objective tallies might help to identify generic
weaknesses in a program, but when used alone often fall
short in offering any remedy. Hence, an instructor may
receive average ratings of 2 (poor) on three of his five
sessions but may receive marks of 9 (good) on the remaining
two sessions. Clearly something is amiss with three
sections; however, simple numerical marks offer no
explanation or possible solution to the problem.

In order to move beyond the identification of generic
weaknesses, one must look carefully at specific comments
provided by participants. Perhaps in the earlier example a
common weakness in visual aids might be suggested by student
comments such as "slides were poorly developed and often out
of focus" or "transparency materials were smudged and not
readable." Such comments provide the functional basis for
the improvement feedback loop. Such comments almost always
identify what the participant perceives as a weakness. Of
course, invalid criticisms are often made and must be
regarded as such.

The feedback loop might also be expanded through
informal discussions between instructors and course
participants. In fact, the suggestion to videotape the Oil
Spill Control Course simulation as a critique came about
partly as a result of such an informal discussion. Advisory
group suggestions often surface in this feedback loop and
emerge as significant course improvements.

In summation, it is necessary to view the four-stage
program development process as a dynamic one. The first
stage of preparation helps to establish a basis for further
development. The second stage involves formulating
objectives and choosing and arranging content material in a
manner to best facilitate learning. The third phase of the
development process involves the actual methodologies
required to translate theory into practice. The fourth
stage concerns analysis of both the program and its

participants. Finally, an improvement feedback loop develops as a logical outgrowth of program evaluation. It is through this link between the evaluation and implementation stages that a program may be updated and improved. The loop itself helps to ensure that the program development process remains dynamic.

WHAT OTHERS ARE DOING

A number of companies have developed excellent training programs, many of which began as in-house projects designed to meet specific corporate needs. One presentation developed along these lines is Texaco's videotaped hazardous materials program.

Developed in 1982, the program was targeted at the multidisciplined, broad-based spill cleanup management teams that had previously undergone Texaco's similarly developed oil spill response training program [5]. The videotape format was selected as the most viable for a number of reasons including cost and in-house availability of existing playback equipment.

The Texaco program appears to be well grounded in its topical approach to subject matter. The program is also well designed in that it makes use of a tutored videotape instruction concept that allows for student-instructor dialogue and employs strategically placed breaks to avoid taxing the attention span of the student. Texaco's premise is that if properly used, the program teaches through repetition [5]. The program appears to be a well-constructed in-house program with enough flexibility for a variety of job levels. While the program lacks significant hands-on activities, Texaco has been careful to point out that this particular program is not directed toward individuals involved in actual hands-on activities relating to hazardous materials incidents.

Other corporate developed courses are Du Pont's "RIT" series and Union Carbide's "H.E.L.P." programs. These programs, similar to Texaco's, were developed in-house for the purpose of training company employees to handle hazardous materials emergencies specific to their industries.

In addition to these types of industry programs, a number of governmental agencies offer training activities. Swiss, et al., describe a response training program developed through a cooperative effort between the Environmental Emergency Division of the Canadian Environmental Protection Service, Atlantic Region and the provincial Environment and Emergency Measures Organization.

This training activity is largely directed at volunteer firefighters and is designed to be presented at various extension centers throughout each province [6].

Leading educational institutions including Iowa State University, The University of Michigan, and Louisiana State University offer various programs in hazardous material and waste control. Other training courses include those offered by cleanup contractors, consultants, and similar groups. Weiss and Leigh in their description of Texaco's in-house training program provided an interesting comparison of hazardous materials control programs offered by a number of groups (see Figure 6).

Such an overview is interesting only as a topical comparison. Curriculum design, methodologies, and philosophical approaches are not covered in this table.

NIOSH developed in 1983-84, an occupational safety and health training program for superfund activities. The three-day course materials concentrate on hazard recognition and hazard control. The courses are available through 15 Educational Resource Centers, Universities that have training grants from NIOSH, located throughout the United States [7].

SOURCES OF TRAINING PROGRAM INFORMATION

With today's emphasis on controlling hazardous materials, wastes, and by-products, a proliferation of training programs aimed at these audiences is hardly unexpected. Many diverse groups are offering programs directed at managing and controlling releases of hazardous materials. Appendix 2, adapted from Volume II of the EPA Directory of Hazardous Materials Response Training, gives a regional listing of the programs being offered and their major points of emphasis [8]. The Directory itself is approximately 190 pages long and gives detailed information about each training program.

Another similar publication, Hazardous Materials - Spills Management Review, prepared for the American Petroleum Institute by the Texas Transportation Institute and the Texas Engineering Extension Service, provides statistical data about hazardous materials training programs. Of those programs documented by this review, 50 percent constitute training courses of some length while the remaining 50 percent are composed of short conferences and seminars. Hands-on training of some type appears in approximately 54 percent of the programs. Most of the programs are oriented toward public safety personnel (38 percent) with a slightly smaller percentage (34 percent)

HAZARDOUS MATERIALS TRAINING[1]

	Texas A&M	National Hzrd. Mtrls. Trng. Program	Contractor	J.T. Baker	National Spill Control School	Center for Professional Advancement	E. I. Du Pont DeNemours & Co. Rhythm Program
Agricultural Chemicals							
Biological Effects					x		
Chemical Treatment	x				x	x	
Chemistry (Properties of Materials)	x	x	x		x		
Chlorine					x		
Communications	x	x	x		x		
Containment	x	x	x	x	x	x	
Corrosives				x			
Cryogenic Liquids-Compressed Gas				x			
Disposal-RCRA	x	x		x	x		
Fate and Effects	x	x		x			x
First Aid			x	x			
Hazardous Mtrls. - Program Overview	x	x	x	x	x	x	x
Health Effects		x		x	x		
Information Systems (CHEMTREC)	x	x	x	x			
Labelling Hazardous Material for shipment				x			x
LPG	x						
Management and Planning	x	x			x		
Material Classification			x	x			x
Monitoring and Detection	x						
PCBs					x		
Protective Clothing and Equipment	x	x	x	x	x		
Public Relations - Safety	x	x	x	x	x	x	
Recovery - Treatment	x	x	x	x	x	x	
Reporting ((EPA - USCG) - Fed. & State Laws	x	x	x	x	x		x
Tests, Inspections	x	x					
Toxicology	x	x		x			
Transportation	x		x				x

1 Information indicated is latest available to the authors.

(Taken from Reference Number 5, Weiss and Leigh)

Figure 6.   Comparison of hazardous material training programs.

319

oriented toward private industry audiences. The remainder (28 percent) of the courses are directed toward governmental personnel and others [9]. These statistics suggest that high priorities are placed upon providing training to both public and private company response personnel. One might conclude that such training would emphasize emergency control and stabilization techniques as opposed to management overviews of hazardous materials. Moreover, the fact that 54 percent of the programs offer some type of hands-on training seems to indicate that programs which are both practical and applicable are favored by a majority of trainees. As in the case of the EPA Directory, the Hazardous Materials - Spills Management Review provides details including course length, topics, tuition costs, and contact person for each program catalogued.

In addition to these two major directories of hazardous material control training programs, there are several other sources of information. Federal agencies such as the Department of Transportation, the Environmental Protection Agency, the National Institute for Occupational Safety and Health, and the Occupational Safety and Health Administration may provide valuable information on program availability. Other groups such as the Chemical Manufacturers' Association, National Environmental Training Association, and the National Tank Truck Carriers Association may also be able to provide information on programs.

CONCLUSION

Today's burgeoning technology brings with it tomorrow's promise of increased production of hazardous materials and wastes. The common challenge facing industry, state and local government, and leading academic institutions is that of ensuring that such technical advances do not occur at the expense of public health and safety.

A practical part of this challenge involves hazardous material and waste control training. Those on the cutting edge of technology must endeavor to see that training should develop in conjunction with emerging technologies. Traditional approaches to training must be critically evaluated and bolstered where necessary with new and perceptive insights. Training techniques cannot remain static; for only through systematic growth and development will training help society rise to meet the challenge of the future.

Appendix I. Typical Course Evaluation Sheet

HAZARDOUS MATERIAL CONTROL COURSE
EVALUATION SHEET

This evaluation sheet will provide important feedback that will allow us to make improvements in the course. Please assist us by responding completely.

1. Circle the appropriate response. Rank only the presentation--not the instructor or topic.

Chemical Properties   1 2 3 4 5 6 7 8 9 10
of Hazardous materials   Poor   Average   Good

Comments:_____
_____
_____

Toxicology   1 2 3 4 5 6 7 8 9 10
                           Poor   Average   Good

Comments:_____
_____
_____

Breathing Apparatus-   1 2 3 4 5 6 7 8 9 10
SCBA Training   Poor   Average   Good

Comments:_____
_____
_____

Hazardous Environment   1 2 3 4 5 6 7 8 9 10
                           Poor   Average   Good

Comments:_____
_____
_____

2. What did you like least?

   _____

   _____

   _____

3. What additional topics or exercises would you like
   to see added?

   _____

   _____

   _____

4. Were staff members and instructors professional,
   well-versed and generally capable?

   _____

   _____

   _____

5. Were visual aids and teaching materials of a
   professional quality, and were they suitable?

   _____

   _____

   _____

	SAFETY	MANAGEMENT	FIELD PROCEDURES	TRANSPORTATION	TECHNICAL	SUPERFUND ORIENTATION	OTHER
ALABAMA							
U.S. Dept. of Defense, Army Corps of Engineers	•	•	•	•		•	
ALASKA							
State Office of Emergency Services	•	•	•	•			
ARIZONA							
Office of Fire Training, Division of Emergency Services	•		•	•			
ARKANSAS							
University of Arkansas for Medical Sciences, Div. of Interdisciplinary Toxicology	•		•	•			
CALIFORNIA							
California Highway Patrol	•	•	•	•			
California Specialized Training Institute	•	•	•	•			
Emcon Associates		•					
U.S. Dept. of Defense, Naval School	•	•	•	•			
University of California, Northern Occupational Health Center	•	•	•	•			
University of California, Southern Occupational Health Center	•	•	•	•			
COLORADO							
American Water Works Association			•				
Colorado Training Institute	•	•	•	•		•	
DELAWARE							
Delaware State Fire School	•	•	•	•			
Delaware Technical and Community College, Fire Protection and Safety Management Depart.	•	•	•	•			
E. I. DuPont De Nemours and Co.	•	•	•	•			
University of Delaware, Dept. of Civil Engineering			•	•			
DISTRICT OF COLUMBIA							
American Society of Civil Engineers		•	•				
American Trucking Association	•	•	•	•		•	
American University, Biology Dept.		•	•	•		•	
Association of Metropolitan Sewerage Agencies		•	•	•			
Chemical Manufacturers' Association	•	•	•				
Hazardous Materials Advisory Council		•				•	
U.S. Dept. of Agriculture, Graduate School	•	•	•	•			•
FLORIDA							
Cross/Tessitore & Associates	•	•	•	•			
Florida State Fire College, Bureau of Fire Standards & Training	•		•	•			
Government Services Institute	•	•		•			
ILLINOIS							
National Safety Council, Safety Training Institute	•	•		•			
U.S. Dept. of Defense, Darcom Ammunition Center	•	•	•	•	•		
U.S. Dept. of Labor, Occupational Safety & Health Administration, OSHA Training Institute	•	•	•	•	•		
IOWA							
Iowa State University, Fire Service Extension		•	•	•			
KANSAS							
University of Kansas, Div. of Continuing Education, Fire Service Training Institute	•	•	•	•			
MARYLAND							
Government Institutes		•				•	
Government Refuse Collection & Disposal Association		•		•			
Hazardous Materials Control Research Institute	•	•	•				

	SAFETY	MANAGEMENT	FIELD PROCEDURES	TRANSPORTATION	TECHNICAL	SUPERFUND ORIENTATION	OTHER
MARYLAND (Continued)							
Johns Hopkins University, School of Hygiene & Public Health, Continuing Education Program	•	•	•		•	•	
Maxima Corp.		•					
U.S. Dept. of Defense, Joint Military Packaging Training Center, Aberdeen Proving Ground	•	•	•	•			
U.S. Federal Emergency Management Agency (FEMA), National Emergency Training Center	•	•	•				•
U.S. Federal Emergency Management Agency (FEMA), National Fire Academy	•	•	•	•		•	
MASSACHUSETTS							
Massachusetts Dept. of Environmental Quality & Engineering	•	•					
Massachusetts Fire Fighting Academy	•	•		•			
National Fire Protection Association	•	•	•				
MINNESOTA							
Department of Public Safety, Division of Emergency Services	•		•	•			
MISSOURI							
D.W. Ryckman & Associates, Inc.	•	•	•	•			
MONTANA							
Montana Dept. of Military Affairs, Disaster & Emergency Services Division	•	•	•	•			
NEBRASKA							
Nebraska Fire Service	•	•	•	•			
NEW HAMPSHIRE							
Environmental Hazards Management Institute	•		•	•			
NEW JERSEY							
Center for Professional Advancement	•			•			
Industrial Training System Corp.	•	•	•	•	•	•	
J.T. Baker Chemical Co.	•	•	•	•			
Lion Technology, Inc.	•	•	•	•			
National Hazards Control Institute	•	•	•				
New Jersey State Fire College, New Jersey State Safety Council	•		•	•			•
Northeast Hazardous Waste Committee	•	•	•				•
Rutgers University, Dept. of Environmental Science, Cook College	•	•	•	•			
U.S. Environmental Protection Agency, Edison Lab	•	•	•	•		•	
NEW MEXICO							
U.S. Dept. of Energy, Sandia National Laboratories			•	•			
New Mexico State Fire Marshal's Office	•	•	•	•			
NEW YORK							
American Institute of Chemical Engineers, United Engineering Center	•	•	•	•			
American Society of Mechanical Engineers	•	•	•	•		•	
Cecos International, Inc.	•	•	•	•		•	
Executive Enterprises, Inc.		•					
Hofstra University, Continuing Engineering Education		•		•			
Ulster County Community College, Water Quality Monitoring Program	•		•	•			
OHIO							
American Industrial Hygiene Association, Institute for Continuing Education	•						
U.S. Dept. of Health & Human Services, National Institute for Occupational Safety & Health (NIOSH), Div. of Training and Manpower Development	•	•	•	•			•

	SAFETY	MANAGEMENT	FIELD PROCEDURES	TRANSPORTATION	TECHNICAL	SUPERFUND ORIENTATION	OTHER
OHIO (Continued)							
U.S. Environmental Protection Agency, Environmental Response Branch	•	•	•	•		•	
University of Cincinnati, Institute of Environmental Health	•		•	•			
Ohio State Fire Marshal's Office, Hazardous Materials Bureau	•	•	•	•	•		
OKLAHOMA							
U.S. Dept. of Transportation, Transportation Safety Institute	•	•	•	•			
University of Oklahoma, Center for Continuing Education	•		•	•			
OREGON							
Environmental Emergency Services Co.	•	•	•	•		•	
Fire Standards and Accreditation Board	•	•	•				
Roberts Environmental Services, Inc.	•	•	•	•			
PENNSYLVANIA							
American Society for Test Materials	•	•	•	•			
Consolidated Rail Corporation (Conrail)	•	•	•	•			
Phoenix Safety Associates	•	•	•	•	•	•	
Transportation Skills Programs, Inc.	•	•	•	•		•	
TENNESSEE							
Environmental & Safety Design (ENSAFE)	•	•		•		•	
Tennessee Military Dept. Div. of Civil Defense, Emergency Operations Center	•	•	•	•			
TEXAS							
Corpus Christi State University, National Spill Control School	•	•	•	•		•	
Texas A&M University System, Texas Engineering Extension Service, Fire Protection Training Div.	•		•	•			
Texas A&M University System, Texas Engineering Extension Service, Oil & Hazardous Material Control Training Div.	•		•	•			
U.S. Dept. of Defense, Air Force, Air Training Command, Sheppard Air Force Base	•	•	•	•			
University of Texas, Continuing Engineering Studies			•	•			
VIRGINIA							
Center for Energy & Environmental Management	•	•	•	•		•	
Darrell Bevis Associates, Inc.	•		•				
U.S. Coast Guard Reserve Training Center, Marine Safety School	•	•	•	•			
U.S. Dept. of Defense, Army Logistics Management Center	•	•	•	•	•		•
WASHINGTON							
Fire Service Training Commission for Vocational Education	•		•	•			
WEST VIRGINIA							
U.S. Dept. of Labor, Mine Safety & Health Administration, National Mine Health & Safety Academy	•	•	•	•	•		•
WISCONSIN							
J.J. Keller & Associates, Inc.	•	•		•			

REFERENCES

1. Taba, Hilda. Curriculum Development: Theory and Practice (San Francisco: Harcourt, Brace and World, 1962).
2. Payne, J. L., and C. B. Strong. "Taking Technology Off the Shelf: Texas A&M's Hazardous Material Control Program," in Proceedings of the 1980 National Conference on Hazardous Material Spills, Louisville, KY, May 13-15, 1980.
3. Mager, Robert F. Preparing Instructional Objectives (Belmont, CA: Fearon Publishers, Lear Siegler, Inc., 1962).
4. Payne, J. L. "Oil Spill Control Training: Texas A&M University's Approach," in Proceedings of the 1977 Oil Spill Conference, New Orleans, March 8-10, 1977.
5. Weiss, H. J. and J. Leigh. "Development of an In-House Hazardous Materials Training Program," in the 1982 Hazardous Material Spills Conference Proceedings, Milwaukee, April 19-22, 1982.
6. Swiss, J. J., W. S. Davis, and R. G. Simmons. "On-scene Response Training Program," in the 1982 Hazardous Material Spills Conference Proceedings, Milwaukee, April 19-22, 1982.
7. Martin, W.F., J.M. Melius, C.A. Cottrill. "Management of Hazardous Wastes and Environmental Emergencies," paper presented at National Conference and Exhibition on Hazardous Wastes and Environmental Emergencies, Houston, TX, March 12-14, 1984.
8. Directory of Hazardous Materials Response Training: Vol. II, U.S. EPA, (Booz-Allen and Hamilton Inc., 1982).
9. Hazardous Materials Spills - Management Review, prepared for the American Petroleum Institute by the Texas Transportation Institute and the Texas Engineering Extension Service, (College Station, Texas: The Texas A&M University System, 1980).

13

CONTINGENCY PLANS

Charles J. Sawyer, CIH, PE
Manager, Environmental Affairs
Syntex Inc.
3401 Hillview Avenue
Palo Alto, California 94304

Uncontrolled hazardous waste sites can present a broad range of potential environmental health and safety problems. Occupational exposure to hazards associated with waste site exploration, sampling, evaluation, and subsequent remediation can be controlled or avoided. The success in controlling adverse exposures depends on the detailed planning, training, scheduling, and execution of a well defined plan to remediate the hazardous wastes site. Such a plan should be drafted utilizing an interdisciplinary team of technical experts, including: Analytical chemistry, geology, hydrology, environmental engineering, industrial hygiene, medicine, toxicology, safety and fire protection, civil engineering, and engineering project management.

A key element of the plan is a detailed, contingency plan. Because of the range of complexities associated with various uncontrolled hazardous waste sites, site specific information must be carefully integrated to develop a satisfactory remedial action/contingency plan. A contingency plan with all its elements is necessary insurance to protect against upset conditions possibly threatening the health and safety of the workers, the environment, and the surrounding environment. A well developed plan should minimize the need to ever effect the callup of a contingency plan. However, if and when an uncontrolled chain of events leads to an emergency situation, a readily adaptable contingency plan with clear responsibilities and sequenced program of activities brought into action should be effective towards prompt restoration of normal operations. Presented in this chapter are those elements that are in part essential to the remediation plan itself but by this focus are necessary to be incorporated for developing a detailed contingency plan. The key

elements for discussion are divided into two categories,
i.e., preventative and emergency requirements.

PREVENTATIVE REQUIREMENTS

The essentials of the uncontrolled hazardous wastes
site contingency plan relative to preventative requirements
are discussed in detail under the following individual
headings.

Know the Inherent Site Characteristics

One of the key steps in the preventative aspects of a
contingency plan is the need for an accurate collection and
evaluation of all known and available information on the
remediation waste site itself. Table I summarizes the
important items needed to evaluate and understand the
inherent site characteristics [1]. Information gaps should
be identified and efforts made to deal with prioritizing
those areas that could most influence the safe conduct of
on-site activities.

Table I. Site Characteristics of Uncontrolled
Hazardous Wastes Sites [1]

Site Characteristics	Related Considerations
Topography	Adjacent tenants
Geology	Nearby population
Hydrology	centers
Climatology	Hospital facilities
Wildlife (reptile,	Ambulance service
animal, insect)	Fire district
Ground cover	Utilities available/
History of sampling/	proximity
exploration	Industrial equipment
Accessibility	rental
Security	Law enforcement

Know the Waste Parameters at the Site

The degree of hazard in remediating the site essentially depends upon the specific waste chemical types, quantities, method of disposal, etc. Table II summarizes key information relative to assessing the waste characteristics at the site [1]. Initially (or at any time) when workers will be venturing into unknown conditions the most conservative protective requirements should be incorporated. If there are blends of wastes, controls

Table II.   Assessment of Key Waste Characteristics
at the Site

---

1. Sources/volume/form
2. History of waste deposits
   *Dates
   *Sources of waste
   *Type and quantity of wastes
3. Sources of additional information
4. Containment/confinement of waste
5. Waste containers
   *Types, age condition
6. Designed confinements
   *Pits, Lagoons, cells, trenches, cover
7. Uncontrolled practices at the site
   *Open dumping
   *Open burning
   *Flooding/evaporation/percolation
8. Hazardous properties of waste site chemicals
   *Physical and chemical properties –
   Flammability, corrosivity, reactivity
   *Potential toxicity –
   Acute, chronic
9. On-site wastes compatibility considerations

---

should be set up to deal with the most toxic chemicals, or the greatest potential hazard, e.g., flammability or explosion. Extreme care is required when dealing with potential incompatibilities of various waste sources at the site.

The conduct of cleanup activities must be under the key control of a single project (site) manager (on scene coordinator) whose day-to-day responsibilities include:

*Monitoring and directing the site activities;
*Planning and scheduling;
*Direct resource management: manpower, materials, equipment; and
*Clear established line of communications to field staff, on-site government agency coordinators, the safety coordinator, as well as press and local government officials.

The need to maintain the highest commitment to safety and health at a site rests with one responsible party who is fully knowledgeable, experienced, and capable of acting with dispatch to monitor the day-to-day activities of personal protective procedures, industrial hygiene sampling and air monitoring, decontamination procedures, weather conditions, etc. The role of the on-site safety and health coordinator clearly is to provide routine advice and counsel to the project (site) manager relative to health and safety matters. Unsafe conduct or disobedience to the documented safety/health procedures serves as a clear reason for ceasing activities until corrective actions are taken.

Training

All cleanup workers must be fully trained and informed on the potential safety hazards at the site, the toxicity parameters of the waste chemicals at the site, protective equipment requirements, decontamination procedures, safe operation of remediation equipment, fire protection, emergency backup, etc. Table III presents pertinent subjects for these workers [2,3,4,5]. Classroom education prior to startup is an effective means to assure adequate worker training. This should include instruction by various technical disciplines, e.g., industrial hygiene, toxicology, etc., as well as trial runs using requisite safety and remediation equipment for practice under controlled, no risk test environments.

The scope and length of the training program should be adjusted to fit the needs of specific worker tasks as well as the complexities and risks of the individual site.

Table III. Training [2,3,4,5]

A. Toxicity of waste site chemicals
1. Acute toxicity
2. Chronic toxicity
3. Dermatologic effects
a. Chloracne
b. Other
4. Epidemiologic studies
5. Other possible health effects
B. Material safety data sheets (MSDS) for major waste site chemicals
C. Potential routes of waste chemical exposure
1. Skin Contact
2. Inhalation
3. Ingestion
D. Respiratory protection
E. Protective clothing/equipment requirements
F. Industrial hygiene and safety requirements
G. Change room requirements
H. Fire fighting techniques
I. Medical monitoring requirements
*First-aid
*CPR
J. Trained to recognize individual medical symptoms possibly indicative of over-exposure to toxic substances:
1. Irritation of skin, eyes, nose, throat or respiratory tract
2. Changes in complexion or skin discoloration
3. Headaches
4. Difficulty in breathing
5. Nausea
6. Dizziness or light-headedness
7. Excessive salivation or drooling
8. Lack or coordination
9. Blurred vision
10. Cramps and/or diarrhea
11. Changes in behavior patterns
K. Standard operating procedures
L. Equipment operation training
M. Decontamination procedures
N. Wastes handling techniques
O. Emergency response plans
P. Hazardous spill control
Q. Personal hygiene and cleanliness
R. Off-site hands-on practice
S. On-site dry runs prior to startup

All training sessions should be attended by the official on-scene representatives of the various Federal, state, and local government agencies.

Personal Protective Systems

The formulation of the preventative aspects of a contingency plan places the highest priority on the protection of the health and safety of the workers. Each phase of cleanup activities must receive a detailed review, focusing primarily on preventing possible exposures to the most toxic chemicals, and secondarily on preventing exposure to other materials of somewhat lower toxicity. Particular attention should be given to the various routes of possible exposure to workers via the respiratory tract, skin, and mucous membranes (eyes, nose, and mouth).

Typically there are four possible categories of personal protective equipment for hazardous material workers [2,3,4]. Selection of specific equipment should reflect the degree of risk associated with specific remediation tasks. The four categories are summarized as follows:

Category I. High Risk for Exposures to Generation of Unknown levels of Aerosols, Dusts, Mists, or Organic Vapor. This set of PPE is similar to Levels A and B as presented in Chapter 9.

Equipment
*Supplied air respirator
*PVC chemical suit
*Chemical gloves taped to suit
*Cotton glove liners
*Neoprene safety boots taped to suit
*Coveralls/underclothing/socks (washed daily)
*Walkie-talkies for communication
*Vortex cooler (for cooling during high ambient
   temperatures)

Category II. Low Risk for Exposures to Generation of Aerosols, Dusts, Mists, or Organic Vapors. This is comparable to Level C in Chapter 9.

Equipment
*Hard hat
*Air purifying respirator with chemical/mechanical
   filters
*PVC chemical suit and chemical gloves taped to suit

*Cotton glove liners
*Neoprene safety boots taped to suit
*Coveralls/underclothing/socks (washed daily)
*Walkie-talkies for communication
*Safety glasses or face shield

Category III. No Risk for Exposures to Generation of
Aerosols, Dusts, Mists, or Organics. (Similar to PPE
Level D in Chapter 9.)

Equipment
*Hard hat
*Disposable coveralls and boot covers
*Lightweight gloves
*Safety shoes
*Coveralls/underclothing/socks (washed daily)
*Safety glasses or face shield

Category IV. Emergency Use. (Either Level A or B in
Chapter 9 depending on protective clothing requirement.)

Equipment
*Positive pressure self-contained breathing apparatus
(SCBA)
*PVC chemical suit and chemical gloves
*Neoprene safety boots (taped if appropriate)
*Coveralls/underclothing/socks (washed daily)
*Walkie-talkies for communication

Category I is the most stringent, providing both
respiratory and skin/mucous membrane protection of workers
engaged in work where a high potential for generation of
potentially toxic levels of aerosols, dusts, mists, or
organic vapors exists. This category also applies when
there exists possible unknown risks or complexities
associated with specific remedial action tasks.

Category II applies to work activities where complete
skin protection is warranted, but the use of air-purifying
respirators is suitable to protect against possible low
level generation of aerosols, dusts, mists, or organic
vapors. The Category II respirator is a combination carbon
filter to remove organic chemical vapors and a HEPA-type
particulate removal mechanical filter.

Category III applies to work activities where no
generation of aerosols, dusts, mists, or organic vapors have
the opportunity to impact upon the worker.

Category IV applies to providing added worker
protection in the case of fire or other emergency.

All respirators must be fit-tested according to
established industrial hygiene practices. Workers wearing

respirators will be trained to assure proper usage, storage, and maintenance. The choice of Categories I to IV will be made after appropriate consultation with the on-site safety and health coordinator. Whenever the risk is unknown, the maximum personal protection of the workers needs to be assured so that Category I requirements are designated.

Medical Programs

All workers and supervisory personnel who are required to handle contaminated materials should be given a comprehensive preemployment physical examination. Table IV summarizes the minimum medical baseline requirements for each employee [2,3,4]. The baseline medical tests should be modified to reflect specific health risks for certain highly toxic chemicals, e.g., dioxin.

Among the list of the key employee medical history and physical examination parameters evaluated by a physician to establish worker preclearance are:

  a. History of nervous disorders, drinking habits.
  b. Evidence of preexisting chronic illnesses.
  c. Skin and liver function evaluation.
  d. Evaluation of suitability to wear respiratory
      protective equipment.

Upon completion of all remedial and closure work, the workers should be given a follow-up physical examination, and another one 6 to 12 months later. This baseline and follow-up documentation reflects appropriate tracking of hazardous waste site workers medical health information.

Site Work Zones

One method for the reduction of possible contamination or release of toxic materials is to class the uncontrolled hazardous waste site into specific delineated work zones or work areas wherein expected or known levels of contamination exist. Within these zones, prescribed operations occur utilizing appropriate protective equipment. Movement between areas should be controlled at specified checkpoints. The three recommended zones are:

Table IV. Health Screening Examination – Minimum
Standards Requirement [2,3,4]

A. MEDICAL HISTORY
   Medical information
   questionnaire acquired
   acquired from
   participant
B. PHYSIOLOGICAL TESTS
   Height
   Weight
   Blood Pressure
     Systolic
     Diastolic
   Vision
     Distance Visual
     Acuity
      Left Eye
      Right Eye
      Both Eyes
     Near Visual
     Acuity
      Left Eye
      Right Eye
      Both Eyes
   Tonometry (Non-
     contact)
   Electrocardiogram
     12 lead
   Audiometry

Right Ear	Left Ear
500 CPS	500 CPS
1000 CPS	1000 CPS
2000 CPS	2000 CPS
4000 CPS	4000 CPS
8000 CPS	8000 CPS

   Chest x-ray
     14" x 17" PA View
   Spirometry
     FVC
     FEV_1

C. HEMATOLOGICAL TESTS
   Hematocrit
   Hemoglobin
   Red Blood Count
   White Blood Count
   Differential (when
     indicated)
   MCH
   MCHC
D. BLOOD CHEMISTRY
   Calcium
   Phosphorus
   BUN
   Creatinine
   BUN/Creatinine
   Uric Acid
   Glucose
   Total Protein
   Albumin
   Albumin/Globulin
   Direct Bilirubin
   Total Bilirubin
   SGOT
   SGPT
   Alkaline Phosphatase
   LDH
   Iron
   Cholesterol
   Sodium
   Magnesium
   Potassium
   Chloride
   GGTP
   Triglycerides
E. URINALYSIS
   Occult Blood
   pH
   Protein
   Glucose

[a]SMA-24 is routine, (standard test conducted by
auto-analyzer).

1. <u>Exclusion Zone</u> – The contaminated area typically
   requiring the most stringent categories of personal
   protective equipment (Category I or II). Within
   this area protective equipment requirements may vary
   slightly based on different levels of contamination
   within the zoned area.
2. <u>Contamination Reduction Zone</u> – An intermediate
   buffer zone between contaminated and uncontaminated
   work areas. (All decontamination activities occur
   in this area.)
3. <u>Noncontaminated or Clean Zone</u> – The outermost area
   of the site where no contamination exists.
   Typically this area contains the bulk of the
   administrative and support services, and serves as
   the focal point for controlled access of authorized
   support personnel and equipment.

Figure 1 represents a typical layout delineating the
three zones [7,8].

Air Monitoring Requirements

A variety of air monitoring equipment is necessary to
characterize and monitor the ambient air at a hazardous
waste site. Air monitoring can document that toxic
materials are not being released at the perimeter boundaries
(both in upwind and downwind directions) during site
activities, as well as to provide information for selecting
the proper respiratory equipment. In addition, continuous
air monitoring is required because workers may encounter
hazards such as explosive atmospheres of high levels of
radiation for which their existing protective equipment
would not be adequate. Such continuous monitoring then
serves as a basis by which the Health and Safety Coordinator
can establish the minimum amount of protective equipment
consistent with maintaining worker health and safety. Table
V provides a detailed list of air monitoring equipment
available for ambient and perimeter boundary air sampling,
and personal breathing zone sampling [8].

Meteorological Monitoring

During all on-site activities, continuous analysis of
the site and vicinity weather is necessary. A portable wind

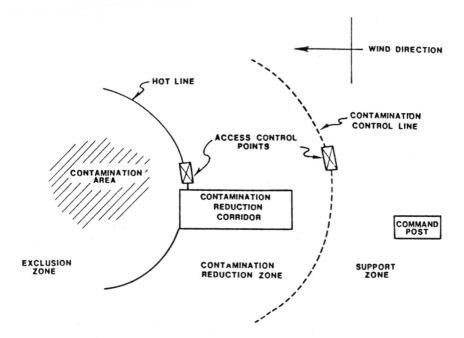

Figure 1.   Site work area classification.

Table V. Ambient/Boundary and Personal Breathing Zone Air
Sampling Equipment [8]

Hazard	Ambient/Boundary: Direct Reading	Collection System
Explosive Atmosphere[a]	Combustible gas indicator	Not used
Oxygen-Deficient atmosphere[b]	Oxygen level meter	Not used
Toxic atmosphere	1. Portable photo-ionization detector (PID) 2. Portable flame ionization detector (FID) w/gas chromatograph (GC) option 3. Colorimetric tubes	Sampling pumps in conjunction with adsorption tubes, filters, and impingers (similar to personal breathing measure-ments)
Radioactivity[c]	1. Radiation survey (alpha, beta, gamma) 2. Passive monitors (alarms)	Dosimeters (film badges)

Table V. Ambient/Boundary and Personal Breathing Zone Air
Sampling Equipment [8]--continued

Pollutant	Personal Breathing Zone: Collection Media[d]	Laboratory Analysis
Volatile organics	Carbon tubes Tenax tubes XAD-2 tubes Silica gel tubes	Gas chromatograph/ mass spectroscopy (GC/MS)
Particulate organics	Gas fiber filters	GC/MS
Pesticides (including PCBs)	Florisil tubes Polyurethane plugs	GC/MS GC/Electron capture
PBBs	Glass fiber filters	GC/MS
Metals	Membrane filters	Atomic absorption (AA)
Volatile inorganics	Impingers/reagent solutions	Wet chemical methods
Particulate inorganics	Membrane filters Glass fiber filters	Wet chemical methods
Cyanides	Filters/impingers	Wet chemicals

[a]At a vapor explosiveness of greater than 35%, all work
activity in the site where reading taken is creased. At
20%, an alert is made to carefully evaluate explosiveness at
various levels in reference to ground level.
[b]At less than 19.5% oxygen, supplied air or SCBA is
required.
[c]0.02MR is normal; at levels greater than 2.0MR,
operations should cease.
[d]Any of these, depending on the specific site hazards, can
be used with a sampling pump to monitor at the uncontrolled
site boundaries for possible off-site release.

station should be established at a point as high as possible
on the site. Continuous recording of prevailing wind speed
(mph) and average wind direction (degrees) is important to
establish locations of upwind and downwind perimeter
boundary air sampling. In addition, wind monitoring will
provide readily available documented wind condition
information by which an uncontrolled release of airborne
toxic materials can be tracked immediately after the
incident. Because of certain instrument calibration
requirements for on-site industrial hygiene monitoring
equipment, it is good practice to routinely record relative
humidity at all times during the site remedial activities.

Security

     As much as possible, the site should be fenced and
isolated from the surrounding population environment. Only
those people approved by the project (site) manager will be
allowed access to the noncontaminated work zone. Only those
workers having completed the employee medical examination
will be allowed into the intermediate and/or contaminated
work zones. Daily reports from the on-site manager should
be made to local government representatives and/or the press
to describe remedial progress and activities. A security
guard service should be employed on a 24-hour basis to
provide on-site security notice of potential exposure
problems, e.g., fire, weather, etc. The guard should
maintain a security log for all workers and visitors
entering the site. The log should indicate name,
organization, dates, and arrival and departure times for
each worker or visitor. At least one or more access routes
for emergency vehicles must be maintained throughout the
site activities. Walkie-talkie communications to remote
field locations are needed to communicate potential hazard
conditions. If at all possible, FAA air space restrictions
over the site at key activity periods should be enforced
using USEPA to facilitate the requirements.

EMERGENCY REQUIREMENTS

     Regardless of the care undertaken to address the major
elements of a remediation plan that can protect (routinely)
against an unforeseen event, there always exists the
potential of an upset condition. The basic objectives for

an emergency contingency plan are threefold: 1) to be prepared to minimize, control, and contain any possible release of hazardous wastes from the remediation site; 2) to provide coordination of all related emergency response groups in a safe manner; and 3) to promote safety in any necessary cleanup operation so as to prevent harm to the workers, the surrounding community, or to the environment. The heart of the documented contingency plan, as identified below, deals with specific emergencies and how each best could be handled [2,3,4].

Fire

 a. Fire extinguishers designed for solvent and electrical fires must be within easy access of various phases of site remediations. (No worker should have to move more than 75 lineal feet to obtain fire extinguishers.) Backup foam systems and emergency fire water supplies should be available as needed to protect against the spread of fire.

 b. All personnel must be trained in the use of the fire extinguishers.

 c. A skilled Fire Brigade must be available, and on call, should a major fire occur. Workers in close contact with the source of the fire should be in Category IV emergency equipment with at least two people participating together.

 d. Telephone (or radio) contact must be made immediately with the local fire department to obtain their response, if necessary, to contain the blaze. Before remedial operations are begun, all fire department personnel should be briefed on the chemical waste hazards at the site.

 e. All site activities will be immediately terminated until the threat of fire hazard is removed. If possible, evacuation from the affected areas should be in an upwind direction.

 f. Electrical power to the work area must be disconnected until the hazard is removed.

 g. All personnel not involved in fire fighting activities should be evacuated to a safe location.

 h. The general plant fire alarm should be activated to notify all personnel of the problem. (Portable air horns will be distributed across the site area for easy access to sound an emergency alarm.)

341

i. All appropriate Federal, state, and local government agencies must be notified of the fire.

j. After the fire has been extinguished, the damage should be immediately assessed and any required spill control or clean-up activities initiated.

Severe Storm

a. At the imminent presence of severe weather conditions, all clean-up activities must cease.

b. Workers should be instructed in proper procedures for securing the site before leaving the work area.

c. As soon as the work area has been secured, all personnel must evacuate the work area, decontaminate, and proceed to a safe area.

d. All electrical power must be disconnected to the work area.

e. All appropriate Federal, state, and local agencies must be notified.

f. As soon as the severe weather passes, the work area should be assessed for damage and any required spill control or clean-up measures initiated.

Toxic Vapor Release

a. Although extremely unlikely, should an unsatisfactory air quality measurement be noted, all site activities must be ceased until the problem is corrected.

b. If necessary, all downwind persons should be notified and evacuated.

c. All appropriate Federal, state, and local agencies must be notified.

d. Personnel in appropriate protective equipment will determine the source of the release and proceed to correct the problem.

Medical Emergency

a. Personnel from local medical facilities must be briefed on the nature of the project so they can make preparations for medical emergencies.

b. Emergency medical transportation must be notified immediately, as well as the nearest hospital.

c. The medical problem should be treated with emergency first aid procedures by qualified trained persons from the site crews until medical help arrives.

d. The affected person, as soon as practicable, will be removed from the work area and protective clothing will be removed, if possible.

e. A designated supervisory person should accompany the person to the hospital to provide the medical team with an accurate description of how the accident or emergency occurred.

Liquid or Solid Wastes Spill

a. Adequate spill control and containment materials must be available on-site to deal with any spill which might occur.

b. All areas subject to potential spills must be diked to prevent migration from the work area off the site boundary.

c. All personnel should be trained in proper spill control measures and must respond immediately to contain, then clean up, the spilled material.

d. All appropriate Federal, state, and local agencies must be notified. Any other unforeseen emergencies should be dealt with by the on-site project manager, and all appropriate agencies will be notified of any additional problems as warranted.

Evacuation

In the unlikely event that the release of toxic materials and/or flammable or explosive mixtures threaten nearby population centers, the following steps should be directed to facilitate the evacuation process:

a. Have a preestablished means to inform quickly nearby population areas. Alternative information contacts include going door-to-door, radio/television broadcasts, or a public address system attached to a motor vehicle driven through the nearby population areas.

b. Have pre-assigned evacuation routes to facilitate transport by private automobile to safe centers wherein temporary food and shelter is available.

c. Secure, monitor, and decontaminate as necessary affected areas prior to any inhabitants returning to their dwellings.

d. Have on file with local authorities, i.e., Civil Defense, National Guard, fire and police departments, EPA, and local hospitals, the evacuation protocol.

e. Have a readily available list of phone numbers and name contacts for emergency situations:

Agency	Telephone	Person to Contact
Fire		
Police		
Ambulance Service		
Hospital Emergency Room		
EPA and EPA Emergency Response Teams		
Mayors of Nearby Communities		
Civil Defense		
Local National Guard Units		
Cleanup Contractor Management Officials		

f. A local communications spokesperson must be established to assure rapid, accurate, thorough, and timely communications to the public during any threat of emergency or evacuation. This is typically the responsibility of the project (site) manager. Joint communications should be drafted and issued among the project (site) manager, USEPA, and local government authorities to assure continuity and consistent appraisal of the emergency status. The utmost care must be taken to prevent a public communications vacuum that can lead to possible panic and misunderstanding to the surrounding populace.

Hazard Risk Assessment

Before startup of any major activity at the uncontrolled hazardous wastes site, an attempt to address the logical consequences of various uncontrolled exposure

344

incidents should be made. Table VI lists several of the
uncontrolled incident cases to which responses can be
drafted in case such emergency situations arise [1]. These
scenario responses should be documented as a key part of any
contingency plan. The response to these issues requires
probable and worst case answers by which careful analysis
and study can identify beforehand important areas of
uncertainty or high hazards.

Table VI. Uncontrolled Hazardous Waste Site
Incident Cases [1]

1. Fire and/or explosion of flammable or combustible
   solvents or pesticide mixtures.
2. Explosion of waste containers containing shock,
   pressure or heat sensitive materials.
3. Penetration or rupture of compressed gas cylinders
   (buried or at the surface) containing toxic
   materials.
4. Penetration of protective gear by toxic liquids,
   gases or vapors.
5. Penetration of protective gear by equipment
   movement, flying debris, or contact with sharp
   objects.
6. Interruption or contamination of supplied breathing
   air.
7. Excavation and surface cave-ins.
8. Equipment rollovers.
9. In transit leaking or rupture of sample containers.
10. Rupture or leakage of sample containers while in
    storage.
11. Violent reaction of waste samples with analytical
    reagents.
12. Medical emergency in hazardous area (e.g., heart
    attack).

CONCLUSION

This chapter has dealt in broad terms with the key
elements of a contingency plan for remediation at
uncontrolled hazardous waste sites. The focus has been to
relate those elements of a contingency plan that are truly
preventative in nature with those that deal with an
emergency situation arising from the remediation activities
at the site. No one uncontrolled hazardous waste site is
exactly like any other. There is no real substitute for a
definitive contingency plan in order to protect workers, the
surrounding community, and the environment. Each
contingency plan needs to be modified reflecting those site
or wastes characteristics unique to a particular situation.
However, the core elements addressed in this chapter must
all be considered preparatory to fine tuning for specialized
hazard situations.

REFERENCES

1. Allcott, G. A., J. V. Messick and R. Vandeewort.
    "Practical Considerations for the Protection of
    Personnel During the Gathering, Transportation, Storage,
    and Analysis of Samples from Hazardous Waste Sites," in
    Management of Uncontrolled Hazardous Waste Sites
    National Conference—October 28 to 30, 1981, Washington,
    D.C., (Silver Spring, Maryland: Hazardous Materials
    Control

2. Sawyer, C. Unpublished material (1984).

3. Sawyer, C. J. "Environmental Health and Safety
    Considerations for a Dioxin Detoxication Process," in
    Detoxication of Hazardous Wastes, J. H. Exner, Ed. (Ann
    Arbor, Michigan: Ann Arbor Science Publishers, 1982),
    pp. 289-297.

4. Sawyer, C. J., and K. E. Stormer. "Environmental Health,
    Safety, and Legal Considerations for the Successful
    Excavation of a Dioxin-Contaminated Hazardous Wastes
    Site," American Chemical Society Meeting-Dioxin
    Symposium, Washington, D.C., (August 1983).

5. Streng, D.R., W.F. Martin, L.P. Wallace, G. Kleiner,
    J. Gift, and D. Weitzman. "Hazardous Waste Sites and
    Hazardous Substance Emergencies," DHHS (NIOSH)
    Publication No. 83-100 (1982), p. 20.

6. Graciano, R., OH Materials Inc., Findlay, Ohio.
    Personal Communication (1983).

7. Partridge, L. J., Arthur D. Little, Inc., Cambridge,
    Massachusetts, Personal Communication (1983).

8. Mathamel, Martin S. "Hazardous Substance Site Ambient
    Air Characterization to Evaluate Entry Term Safety," in
    Management of Uncontrolled Hazardous Waste Sites
    National Conference-October 28 to 30, 1981, Washington,
    D.C., (Silver Spring, Maryland: Hazardous Materials
    Control Research Institute, 1981, pp. 281-184.

347

Aliphatic amines,
   air monitoring method, 110
Ammonia gas,
   respirator selection for, 187
Anomalous permeation, 200
APER: Air Pollution Emission Report, 74
API: American Petroleum Institute, 304, 318
APTIC, 76
Aquifers, 66
Asbestos,
   air monitoring method, 110
ASI: American Statistical Index, 76
Asthma, 148
Atomic Energy Act, 18

Banana oil, 195
Barrel bulking, 6, 129, 164, 166
   chamber, 176
   temperature monitors, 140
Barrel grappler, 6, 175, 176
Barrel handling, 164
Barrel opening, 117
   air monitoring during, 122
   punches, 6, 172, 176
   remote, 166, 172
Barrel overpacks, 172
Barrel staging, 174
Barriers, 160, 175
Bases,
   as decontamination solvent, 271
   field study air monitoring results, 122
   compatibility testing for, 129-145
Benzaldehyde, 142
Berms, 175
Bid specifications, 3
Black body sources, 216
Blood samples, 5
Biological monitoring, 149, 150
Biosis Previews, 76
Bobcats, 175
Botsball, 244
Breakthrough time, 198, 200, 203
Breathing zone, 107, 114, 122
Bulking, (See Barrel bulking)

CAASE: Computer Assisted Area Source Emissions Gridding
   System, 74
CA Condensates '70-71, 76
CA Condensates/CASIA, 76
CAB Abstracts, 76
Cancerline, (See Cancerlit)
Cancerlit, 61, 76
Carbon monoxide,
   respirator selection for, 187
Carbon tetrachloride, 138
   as decontamination solvent, 271
Cardiopulmonary resuscitation, 46

350

FMSHA: Federal Mine Safety and Health Act, 18
Forensic Services Directory, 61
Forklifts, 175
Fountain Avenue Landfill,
    toluene air levels at, 123
Freedom of Information Act, 15
FRSS: Federal Register Search System, 62
FT-IR: Fourier Transform Infrared Spectroscopy, 5
    analysis of ambient air, 121
    analysis of bulk waste samples, 121

Gamma survey instrument, 133, 339
Gas collection system, 178
GC: Gas chromatography, 122, 137, 203, 339
GC-MS: Gas Chromatography - Mass Spectrometry, 5, 129, 339
GEGU, 218
Generator,
    records of, 64
Geoarchive, 82
Geology, 66
    hydrology, 66
Georef, 82
Globe temperature, 223, 233
Gloves, 204
Grappler (See Barrel grappler)
Groundwater,
    affected by climate, 65
Government programs, 15-34

HACS: Hazard Assessment Computer System, 28
Hardhats, 184
HATREMS: Hazardous and Trace Substances Emissions Systems, 74
Hazardline, 62
Hazardous Material Control Course Oversight Committee, 304
Hazardous Material Control Training Program, 7
Hazardous Materials Spills Management Review, 318
Hazardous Waste Site Identification and Preliminary Assesment
    form, 68
Hazardous Waste Site Survey Record, 68
Health and Safety Programs, 35-55
    air monitoring, 106
    benefits of, 53
    budget, 51
    committee, 39
    cost of, 38
    employee involvement, 39
    goals of, 37
    negative economics, 38
    program director, 51
    risk evaluation, 40
    staff selection, 50
Health hazard evaluation, 19
    first strategy, 2
Health Surveillance System, 41-44
    components of, 43
Heart rate, 253

physical state effects, 109
repetitive, 115
selecting contaminants to monitor, 108
verify site control procedures, 170
SANSS: Sub-structure and Nomenclature Search System, 62
SAROAD: Storage and Retrieval of Aerometric Data, 74
SCBA: Self-contained breathing apparatus, 5, 104, 191, 192
protection factor for, 190, 191
Scisearch, 88
SDRITS: Simultaneous direct-reading indicator tube system, 119
SDS Search Systems, 60
address, 60
SDWA: Safe Drinking Water Act, 59
Security, 160, 168, 340
site headquarters, 169
Sensitive environments, 67
SFFAS: Superfund Financial Assessment System, 63
Shepard's citations, 61
Shock-sensitive containers, 141
SIEFA: Source Inventory and Emission Factor Analysis, 74
SIPS: State Implementation Plans, 74
Site inspection, 68
Site investigator,
initial, 3, 68, 72, 160
Site Investigator:
health and safety of, 48, 72
training of, 72
Site layout, 159-182
Site map, 160
working areas, 160
Site preparation, 174
Site ranking, 3
Sketch map, 71
Skin adsorption, 192
Skin temperature, 261
convergence, 256
tolerance times, 261
Smoke detectors, 177
Sodium, 135
Soil, 65
Solid sorbents, 109
Solvent exposure, 150
SOTDAT: Source Test Data System, 74
Sparkless tools, (See Barrel punches)
Specific heat of body, 215
Splash shields, 175
Splash suits, 185
Storet, 88
Sulfide, 130
air monitoring method, 119, 140
generation of gas, 139
in wastes, 138
Sulfur, 138
Superfund (See CERCLA)
Support zone, 161, 168, 337
personal protection levels in, 168
Surfaces,